DATE DUE

DEMCO 38-296

MICROCOMPUTERS ON THE FARM

SECOND EDITION

Microcomputers on the Farm: Getting Started

SECOND EDITION

DUANE E. ERICKSON
ROYCE A. HINTON
RONALD D. SZOKE

IOWA STATE UNIVERSITY PRESS / AMES

Duane E. Erickson is Professor of Agricultural Economics and Farm Management; **Royce A. Hinton** is Professor of Farm Management and Extension Specialist in Farm Management; **Ronald D. Szoke** is Research Programmer and User Training Coordinator, Computing Services Office; all at the University of Illinois, Champaign-Urbana.

First edition, 1985
Second edition, 1990

Library of Congress Cataloging-in-Publication Data

Erickson, Duane., 1931-
    Microcomputers on the farm: getting started/Duane E. Erickson, Royce A. Hinton, Ronald D. Szoke. – 2nd ed.
        p.   cm.
    Includes index.
    ISBN 0-8138-1157-0 (alk. paper)
    1. Farm management–Data processing. 2. Microcomputers.
I. Hinton, Royce A.   II. Szoke, Ronald D., 1934-   . III. Title
S565.7.E75 1990
630′.68–dc20

CONTENTS

PREFACE

This book presents the essentials of acquiring and using a small computer (microcomputer) in agricultural operations. It can serve as a guidebook; it covers background information about microcomputers and how they work, discusses how they can be used by farmers and ranchers, and includes a glossary of common computer terminology.

New technology is increasingly necessary as agricultural operations continue to grow in size and complexity. The three decades of the 1960s, 1970s, and 1980s have shown dramatic changes in the various sales classes of farms in the United States. The total number of farms is declining, but farms are increasing in size when measured in gross sales. The number of farms with sales over $40,000 has increased over five times in the last three decades, from 110,000 to 566,000 between 1960 and 1987, according to the USDA (Agricultural Statistics Board, NASS, USDA, A-31, August 1988).

Farms with higher gross sales need information systems that allow for better decision making. The planning, directing, and controlling of farm businesses has become even more critical in the 1990s with the higher interest rates, higher costs of production, and large variations in prices of products sold. Currently, agriculture includes more business risks; this means that farmers must obtain timely information in order to make decisions. Microcomputers can be used to organize information for accounting, inventory control, and planning.

Farm management can be defined as the application of agricultural techniques and economic principles to the organization and operation of a farm to secure the greatest net return, which may be in terms of maximum satisfaction or maximum financial gain. A number of management functions are performed by most farm operators. The five major management functions are planning, organizing, directing, coordinating, and controlling. All these functions must be considered when choosing the appropriate information system.

MICROCOMPUTERS ON THE FARM

SECOND EDITION

First Steps

Will you be one of tomorrow's successful farmers? A modern commercial farming business involves the use of considerable amounts of land, capital, equipment, supplies, and labor in the production of agricultural products to sell at a profit. A successful farm operator obtains and organizes these resources, makes decisions about them, and takes action to use them most effectively to attain certain goals. To be successful, you need skills in observation and logic and in acquiring information and using it. That's where a microcomputer comes in. Computers are useful in storing items of data, retrieving them when desired, and performing numerical calculations.

The usefulness of the computer in improving managerial skills depends directly on the farm information system. Accumulated farm financial transaction data is not necessarily information. Data becomes information when it is effectively used in planning, decision making, or carrying out decisions. Therefore, in evaluating the use of a computer in your farm business, the focus should be first on the choices you need to make in planning, directing, and coordinating your farm business and on the information you require to perform these tasks. Second, you must consider the alternative sources of computer hardware and software in terms of what information they can provide. In other words, you should list the choices you make and the tasks you perform, then list the kinds of information needed for each choice or task, and finally consider the potential sources of this information.

INFORMATION NEEDS

Extensive information is needed for performing the various farm managerial tasks. Some of the data comes from internal sources (farm records), while other data must be acquired from external sources. In the day-to-day coordination and administraticn procedures of a farm operator, most of the needed data comes from internal sources.

Examples of reports generated from internally produced data are billings and payroll checks, production and sales schedules, income tax statements, partnership reports, true profit statements, balance sheets, and enterprise performance records. Appropriate original record documents listing quantities, rates, prices, etc., are needed for preparing these statements and reports. For any report, however, a systematic method of verifying the accuracy and completeness of information is also necessary.

A broad base of information sources is required for managerial planning and decision making. Planning requires identification of the problem and then the information needed for deciding on one of the alternative ways to solve it. Part of the planning base must be your basic status-of-the-business reports, namely, the balance sheet, the true farm profit statement, and enterprise production performance records.

How financially well-off are you? Your balance sheet and income statement can tell you. Analysis of your balance sheet using various liquidity and solvency measures such as current ratio, debt structure rates, leverage ratio, and net capital ratio can reveal major financial problems; analysis of your income statements can reveal factors influencing your net earnings.

To succeed in farming, you must be an efficient producer and a strategic marketer as well. Efficient producers are those who achieve low costs by producing high yields (bushels per acre, milk per cow, or pigs per litter) while holding production costs down. Finding out how you stand on production efficiency means comparing yields and costs with a standard such as state averages for operations of a similar size and type. Or better yet, you can rate your performance by comparing it with your own goals.

Finding out where you stand on any performance measure calls for keeping and analyzing both total farm records and individual enterprise records. This can be quite a chore, but you have to know where you are before you can chart a course for where you want to go.

DECISION AIDS

After deciding what information you need to manage your operation effectively, you should search out sources of this information. Much of this will come from sources off the farm. Farm magazines; extension circulars; and farm machinery, fertilizer, chemical, and fuel dealers can be sources of information on costs and prices of buildings and equipment. For decisions regarding what crops to grow or livestock to raise, any physical-yield and cost information from previous years will be a help. In addition, off-farm sources of predictions of future prices and costs offer important planning data. Input forms with a detailed list of questions for software programs that perform calculations are tools that can help you to decide what to produce.

Software programs also can help identify the factors you should consider when making a choice between alternative courses of action. For example, choices between equipment investment alternatives depend not only on profit but also on your projected cash flows and their timing. Projected cash flows for the total farm are essential for planning by farm operators using borrowed funds in their operations.

SOFTWARE AND HARDWARE

Understanding the terms "software" and "hardware" is essential before beginning an evaluation of how useful a microcomputer would be in your operation.

Software is the set of instructions that tells the

microcomputer what to do. It is the information that makes the computer operate and perform various functions. The best-known example of software is a **program** (a list of detailed coded instructions) that tells the computer how to carry out some information processing procedure. Two types of software (discussed in detail in Chapter 6) are operating-system software and applications software. The operating-system software, which is specific to each microcomputer, allows the manipulation of files containing data and the use of programs with utility operations, such as saving, loading, copying, printing, editing, deleting, and merging data as well as running individual software-application programs. Software-applications programs allow the microcomputer user to do bookkeeping, inventory control, and planning tasks. (See Chapter 5 for a more thorough discussion of software and the Glossary for definitions of computer terminology.)

One way of getting started in deciding on the value of a microcomputer for your operation is to check out the available software programs for the single reports you are now doing by hand. Some experts consider this step too elementary in light of the complex functions a microcomputer can perform. But it can be valuable for anyone unfamiliar with microcomputers and skeptical about their usefulness.

Hardware is the term used for the physical parts of the computer. The central processing unit is the main control and mathematical unit. The monitor (screen), keyboard, printer, cables, and modems are peripheral hardware components. A more detailed discussion of this hardware will be found in Chapter 6.

STAGES OF MICROCOMPUTER ADOPTION

Anyone who adopts a microcomputer passes through a number of stages. These stages, which are found in any technology adoption, include (1) awareness, (2) interest, (3) trial, (4) evaluation, and (5) acceptance or rejection (the decision stage).

The question that farm operators raise is, How do I decide whether a microcomputer is for me? In general, you can begin to answer this question during the *awareness stage* by assembling information in a three-ring notebook. Gather this information by visiting computer stores and any meetings or exhibits sponsored by a university or community college, the Cooperative Extension Service, or commercial interests where computer information is featured.

During the *interest stage,* you should explore your essential information needs. Ask yourself what financial production cost data and other performance information are required to effectively and efficiently run your farm or ranch. What periodic reports are needed on income and expenses, balance sheets, and cash flows? Who needs reports, the operator or landlord? Are enterprise performance reports needed? What labor information is needed? Are these reports needed only for this year or in coming years?

What decision-making aids would make the operation run better? Determine the business decisions that you make daily, monthly, annually, or even only once in a lifetime.

The *trial stage* is an exciting one. During this time you should obtain some practical, hands-on experience with a microcomputer. A high school, community college, junior college, university, computer store, computer-owning neighbor or family member, or software dealer may offer the opportunity for such experience. Remember, this also is the time to consider who might run the microcomputer if you purchase one. A family member with particular skills may be the logical choice as the major microcomputer operator, but the major decision maker or decision-making group should also be able to operate the chosen computer, using the appropriate software programs. The trial stage is the same whether shopping for a computer or for a tractor, automobile, farm equipment, or household equipment and devices; family members who will be using the purchased item the most should be heavily involved in trying out the models.

The *evaluation stage* is particularly important in both hardware and software purchases. Software evaluation should be completed before or along with hardware evaluation. Family members and perhaps other members of the farm or ranch operation should be involved at this stage. Decisions must be made about who will be doing the data entry jobs and who will be using the reports that are generated. Detailed evaluation of the available application software programs by the farm or ranch firm members who will use them is essential. Although some family and firm members may at first be hesitant about getting involved, some of the best evaluations can be made by experienced farm or ranch operators, wives, daughters, sons, and hired accountants, who are aware of the needs of a particular operation. Trying out a new product takes time, of course. Most farm operators have slack periods when this doesn't interfere with other activities. *But be sure to take your time!* If no slack time is available, make time. Chapters 5 and 6 have sections that deal with the evaluation of microcomputer software and hardware, and the checklists in Chapter 7 will be helpful.

The final stage of the adoption process is acceptance or rejection, the *decision stage*. The advantages and disadvantages of both software and hardware for your operation must be listed. A final decision should be based on this information.

ASSEMBLING INFORMATION ON HARDWARE AND SOFTWARE

The stages in the adoption process should provide you with a general idea of whether a microcomputer is for you.

If you decide that a microcomputer will be an asset in your operation, begin your selection process by assembling information on available microcomputer hardware and software. Check the schedules of local educational institutions for classes, seminars, and demonstrations on microcomputers. Set up a schedule of meetings you will want to attend; most farm families take from three months to a year in this information-collecting stage. Remember to use the three-ring notebook with dividers to store the information you collect. (You may find some contradictions in the information collected, since it depends on particular hardware and software combinations.) The assembling of information in an organized manner, listing advantages and disadvantages, will save time and make your information assembly process more effective.

CHOOSING SOFTWARE

Discussing and settling software acquisition questions before purchasing hardware is a *must* in the interest, trial, evaluation, and adoption stages, since software selection is the key to successful operation of a microcomputer. The following suggestions may be useful.

The first step in the process of software selection is observing a demonstration of a particular software-applications program. Start with a simple program. Then go to a more detailed program or package, such as one for farm records. The usual source for this is a commercial software vendor. Lists of firms with this type of package can be found in a commercial microcomputer software source guide. Once you have located a program that sounds like it will do what you want, find a computer dealer who sells that particular package and ask for a demonstration. (One way to test the program is to use data from previous operating records to check the quality of data entry.)

All software packages should have user guides, which are sometimes referred to as **documentation.** A second set of documentation includes the specific language in which the program or package is written and the detailed flow chart, the blueprint of the computer program construction. Study the user guide documentation. It should have an overview, a tutorial section, a summary of commands, a troubleshooting guide, a technical section, and a table of contents or index. More details on these items are included in Chapter 5.

Evaluate hardware and software supplier (vendor) support. (A vendor may supply hardware or software or both.) A vendor should be willing to provide support after you have made the purchase. Study the computer dealer or vendor as though you were selecting a local farm machinery dealer. Does the vendor answer telephone calls on software problems? Does the vendor have written warranties as well as low-cost backup copies and/or updated copies of software if something goes wrong with your copy? Does the vendor supply initial on-farm or seminar training in the use of software? What kind of service can be expected if the software doesn't work? Does the vendor conduct seminars and provide newsletters?

And finally, try out the particular software package (hands-on experience) before making the purchase. This is the most important part of the trial and evaluation stages. A number of questions can be answered in the hands-on stage. Does the software package have menus or prompts (see Chapter 5 and the Glossary) to help you run it? How does the software-application package or program handle errors? How accurate is the program? How fast does the program run? What printout options are available? How easy is it to use? Can the package or program be modified? How useful does the software-applications package or program seem to be for your farm or ranch operation? A detailed discussion of each of these questions can be found in Chapter 5.

A STUDY OF MICROCOMPUTER PURCHASES

Twelve Illinois farm families who had purchased microcomputers were interviewed by telephone to learn the processes they went through before making the purchase (see Chapter 3 for details). They were asked, How did you go about buying the microcomputer? These families had purchased their equipment between 1977 and 1988. Five had looked for one year, five had looked for two years, one had looked for 3 years, and one had looked for four years before purchasing a microcomputer. All twelve had studied software for bookkeeping or farm accounting before making the purchase. Eleven had attended University of Illinois Cooperative Extension Service meetings. Three had taken community college classes before making the purchase of hardware or software. Three of the twelve felt that consultant services were most important in getting the maximum benefit from their hands-on experience. All twelve indicated they had attended commercial seminars on microcomputer use. (See Table 1.1 for a compilation of the responses to the interview questions.)

SOCIAL CONSEQUENCES OF OWNING A MICROCOMPUTER

The social consequences are seldom considered by farm families or farm operators when they decide to purchase a microcomputer. For example, the person who is learning how to run the microcomputer may initially spend a large amount of time mastering the use of the software programs and tend to ignore other members of the family during the learning process and even later. Time spent with the microcomputer is in competition with time spent in other family activities. Individual family members should discuss these social consequences; a certain amount of patience and tolerance from everyone is needed during the learning phase of operating a microcomputer.

Another social consequence of acquiring a microcomputer is related to the control of information. The person who enters data associated with the farm accounts, such as the posting of receipts and expenses, gains knowledge of the business, which can place that individual in a position of power. Therefore, husband and wife management teams may benefit from both periodically operating the computer and entering data; a son, daughter, son-in-law, or daughter-in-law can learn about the overall business operation, financial growth, and financial stability by having access to the use of bookkeeping, inventory control, and planning-applications programs. Family members should take care to discuss the computer-generated reports so that joint decision making is possible.

A learning potential exists for all family members having access to the microcomputer. Family members who are in

Table 1.1. Survey of Twelve Farm Families Purchasing Microcomputers

	Family											
	1	2	3	4	5	6	7	8	9	10	11	12
Year computer purchased	1980	1977	1983	1980	1984	1983	1987	1988	1983	1983	1985	1988
Years spent looking	1	1	1	2	2	1	3	4	1	2	2	2
Studied software and hardware together	X	X	X				X	X	X	X	X	X
Studied software												
Bookkeeping	X	X	X	X	X	X	X	X	X	X	X	X
Marketing information		X	X				X	X	X	X	X	
Forward planning							X	X	X	X	X	
Studied hardware and looked for software		X				X						X
Studied software and hardware with help of consultant			X	X	X							
Made detailed cost analysis after studying software and hardware					X		X		X	X	X	
Purchased computer after viewing it at trade show												
Took community college courses				X	X			X				
Attended University of Illinois Cooperative Extension Service meetings	X	X	X	X	X	X	X	X	X	X	X	X
Attended other commercial seminars	X	X	X	X	X	X	X	X	X	X	X	X

elementary school, high school, or college may benefit from programmed-learning software-applications programs for basic skills in mathematics, languages, and other subjects. Of course, it may be necessary to schedule microcomputer time to ensure equal access to its use.

Farm Accounting and Other Farm Record Programs

Let us assume that you are a crop farmer and have a modest record keeping system. You maintain a single-entry financial record of income, expenses, and capital accounts. You keep a notebook that carries field records of crop yields, soil tests, fertilizer treatments, seed varieties, and pesticide use by land ownership tract and another notebook to keep an inventory of grain sales records. You also have a record in the back of your farm account book of moneys borrowed and the repayment schedule of interest and principal for current, intermediate, and long-term borrowing. All of these records are kept on an annual basis and filed. This chapter discusses how a microcomputer can be used to handle such record keeping.

CASH AND GENERAL LEDGER PROGRAM

The first program you might examine is an accounting package. There are two different types of computerized accounting systems available: cash and general ledger systems.

A cash system allows entry of cash receipt and expenditure transactions. Only revenue and expense accounts are contained within the system, thereby allowing only the direct production

Much of this chapter is based on results of research reported in *Accounting Systems for Farms: Design and Selection Method* by Gary Schnitkey, unpublished University of Illinois M.S. thesis, August 1984, and in *Computerization of Crop Production Records* by Roy Delaine Wendte, unpublished M.S. thesis, University of Illinois, August 1984.

of cash flow reports. Most cash systems have facilities for computer preparation of income statements, namely, adding the inventory depreciation accounts. With the cash system, balance sheets are generated independent of the accounting system.

A general ledger system includes asset, liability, owner's equity, revenue, and expense accounts. When entering data, each transaction has two entries that affect different accounts and balance each other. In some systems the second entry is automatic; in others it is not. At all times the accounting equation of assets plus expenses must equal the sum of liabilities, owner's equity, and revenues. Balance sheets and resource reports can be produced directly from the general ledger system.

The type of reports the accounting system can generate is an important factor in judging a software package. A simplified accounting system may generate the following classes of reports:

Transaction entry reports. Entry reports summarize all inputs (entries) into the accounting system. This allows for data validation and a track for determining account balances.

Total farm reports. These reports summarize the operations profitability (income statement), cash flow, and sources and uses of funds and collect data for completing Schedule 1040 F income tax liability.

Ownership split reports. When a farm has multiple owners, split-ownership reports can list each owner's receipts, revenues, payments, and expenses.

Responsibility center reports. These reports detail the financial records of individual segments and/or enterprises of the business.

How frequently a computer can generate reports may also be a factor. Yearly, quarterly, monthly, or other user-defined periods are common in computerized system options. The possibility of budget comparisons and/or comparison with last year's operation may be an important factor in deciding whether a computer program will be useful.

The data in the reports that are generated are determined by the number of characteristics included in the transaction entry. The input for an individual entry can vary from two to six items, which may include date, description, reference number, quantity, amount of transaction, and account affected. A cash system has only one entry per transaction, whereas a general ledger has two entries per transaction.

Two methods of inputting information are available. One is the manual entry of all information and the other involves computerized check writing. In the manual system the information is transcribed from checks, receipts, or other source documents. In the other system a check writer prints a check as transaction data is entered into the accounting system. Because much of the time involved in using an accounting system is spent entering transactions, you should consider ease and speed of inputting and validating entries. In addition, because errors are often made in entering data, ease of finding and correcting errors is another criterion to consider when you judge an accounting system.

TRANSACTION SUMMARIZATION CRITERIA

An evaluation of computerized account systems can be made on the basis of the way the summarization process can be tailored to individual farm needs. One factor concerns the accounts. Items for you to consider include the number of accounts allowed, availability of user-defined account names or numbers, subdivision into subsidiary accounts, how to make changes in account name, and how to add new accounts or delete old ones at a later time.

Another factor concerns the handling of split ownership and the number of splits allowed. To some extent, this is related to subdivisions of a farm business.

Enterprise accounts can generate reports of business-segment income and expenses. Factors to consider include the divisions permitted. Are they user defined? Can personal account transactions be separated from business transactions? Are subsequent changes in enterprises or responsibility centers allowed?

Some accounting systems allow more than one checking account so that reconciliation of each account can be made. Subsidiary ledgers and registers give more information on specific assets or liability accounts. These are separate on cash system accounts. Only on general ledger systems are these accounts integrated. Common subsidiary ledgers or registers include accounts receivable, accounts payable, inventory, fixed assets, payroll, and loans.

One set of transaction criteria deals with processing:

Period processed. How often are the summaries made? The options are yearly, quarterly, monthly, or user-defined periods.

Speed of processing. How quickly does the software complete the transaction summary?

Storing of account balance. For what length of time are the totals from the procedure stored?

OTHER DATA FILE FARM RECORD REPORTS

In addition to organizing farm accounting data, the computer can assist farm operators in maintaining records of physical materials use, inventory records, and production records for various farm enterprises or responsibility centers as well as for personal records. The computer programs for records can be either an organized, special purpose one or a flexible data base management package.

The organized information record and summary programs are structured to record certain items of information and generate specific types of summary reports. They are generally simpler for the beginner. The data base management packages have the advantage of flexibility, allowing you to arrange your own data base and determine the reports to be produced. These simple packages allow file definition, data programming, sorting, and creating reports. Without programming knowledge you may easily organize and retrieve information about your business. First, you designate file names and entry fields. After entering data, you may sort for records that meet certain criteria and then print a report of them. Sizes of systems vary, but the limitations of any of them become apparent when more than one file is needed to generate a report. Most farmers who wish to keep and use detailed records on crops and other enterprises find that the simple file management packages fall short of their needs. They are best suited for keeping mailing lists and inventories.

The more comprehensive data base systems are not easy for the beginner to master. To use them, the operator must learn the language of the program to create and define files and, in addition, must learn the programming commands to put the files together and generate multiple reports.

In evaluating data base management packages for farm applications, look at the configuration and structure of the data base, input and output functions, update capabilities, report-generation features, inquiry function, performance, and ease of use. In addition, look at the company, its support, and its documentation.

When choosing the program package to use for computerizing production and other farm performance data you need for management of your farm operation, there are additional factors to consider. The system should allow you to expand and keep track of additional items that are not presently in your hand system. Calculations within the data base should be possible. Ideally, since you may change your mind or situations may change, reports should be flexible. Also, there should be convenient ways for you to handle problems associated with missing data or incompatibility of data within and among files.

CROP ENTERPRISE RECORDS

No doubt you are already recording selected items of information on crop production, quantities of fertilizer used, and pesticides applied to each field. Entering such data into a computer and adding new information as it comes along may be a challenge at first. Ultimately, however, you will be able to use the information you have recorded, such as field records, in decision making. Field record identification data include the name, field coordinates, and owner. Descriptive data include the number of acres, soil type, yield potential for each crop, soil test results, and test dates. Annual physical input and output data include such items as the quantity and quality of production and the time and quantity of input applications (e.g., fertilizer, pesticides, irrigation, and even machine operations performed). Other field data that you could record are scouting reports on plant population and insect and other pest populations.

Other crop enterprise data that would be useful to computerize are items such as enterprise costs, marketing sales, records, crop inventories, rainfall and temperature records, and degree-days.

Ideally, these crop records should be summarized in a format that would help you make decisions regarding what crops to grow and what practices to follow in the various fields. At the present time there are no software programs that allow direct integration of field records into decision models. Users must make manual transfers of data into available decision aid models.

LIVESTOCK ENTERPRISE RECORDS

Production records for livestock enterprises should be kept for management purposes. Every item recorded with your hand system can be monitored with a computerized record system. Many farm operators expand the items recorded when a computer is used because little effort is required to summarize the additional numbers. Ideally, you should record and generate reports only for the data needed for your management decisions.

Production performances by individual animals and for the total unit are the most frequently recorded statistics (e.g., milk produced, weight gained, animals born, animals died). Status of breeding animals is another usual category (the date bred, pregnancy test date, and results), and similar items can generate dates of expected calving or farrowing and time for needed treatments.

Recording data on feed use and weight gain will provide data for comparison of feed conversion between units and for analysis of feed costs and weight gain efficiency. Other data that are useful to computerize are records of nonfeed production items, market sales, treatments, and environmental factors such as temperature and humidity.

Computerized production records for livestock can be developed from the data base management system packages or from specific programmed record and summary programs. Selection should be based on usefulness of the program. In some cases, such as custom feeding of livestock, your requirements for billing customers for selected feeds and services may require a program written for your specific needs.

As in the case of crop records, there can be no direct inputting of historical record summary data into computer decision management programs for livestock. Data for least-cost rations, break-even prices, and planning programs must be manually entered.

Experiences of Farm Microcomputer Owners

The following experiences of 12 families who purchased microcomputers illustrate the many ways farmers and ranchers are using them for bookkeeping, marketing, field records, livestock analysis, cash flows, and communications. (You will find definitions and detailed discussions of computer terminology in Chapters 5 and 6 and the Glossary.)

Lowell Tison, Jr., a grain farmer in Eldorado, Illinois, has used a microcomputer since 1977 when he purchased a Radio Shack Model I. He later traded his Model I for a Radio Shack Model 4 when it became available. He is a cooperator in the Farm Business Farm Management Service of the University of Illinois Cooperative Extension Service. The commercial software package he selected, a farm accounting program, was compatible with the annual summaries created in the record keeping service. The farm accounting package has a check writing program that allows allocation and retrieval of expense information by local agribusiness firms.

Lowell operates 670 acres of owned land and 300 acres of share-rented land that is planted in corn, soybeans, wheat, and hay crops. He farms jointly with his brother, Joe Tison. For income tax purposes the two operators keep separate accounts. They also calculate the costs of production for the various crops.

The system Lowell purchased includes 128K of random-access memory, a printer, a monitor with a high-resolution screen, and a Novation CAT 300 modem. He uses the modem to connect with a commercial marketing information and advisory service, AgriData Network.

Business reports that he generates by using the microcomputer software program include financial accounts and records, annual cash flows, bimonthly cash flows, Internal Revenue Service Schedule 1040 F tax information, loan activity summary, family living expenses, enterprise profit-and-loss statements, and check writing. Lowell considers all of these reports very useful in his farm operation.

Lowell indicated that keeping the farm records was the primary use of the microcomputer on his 970-acre farm. The check writer program associated with the SECRETARY OF AGRICULTURE farm record program (now TRANSACTION or TRANSACTION PLUS) purchased from Farm Business Systems of Aledo, Illinois, allows him to itemize expenses for the various crops grown. Obtaining the most current information is a first step in making better decisions during the year. Projections of cash flow are also easier to develop if cost estimates are available. The printed summaries from Farm Business Farm Management record analysis are also helpful in the process of assembling costs.

In addition to farm accounting, the Tison family uses its microcomputer for other purposes. They have 10 educational software programs that are used by various Tison family members.

Steve Wentworth, a young farmer in central Illinois, became interested in microcomputers in 1979. Over a one-year period he attended several University of Illinois Cooperative Extension Service meetings, visited computer stores, and gained hands-on experience on a microcomputer. Steve's wife, Pam, also was very much involved in obtaining information during the awareness stage. Prior to the microcomputer purchase they studied University of Illinois Department of Agriculture Economics staff paper 81-E200, "Thinking about a Microcomputer." After reviewing a number of software and hardware combinations, Main Street Computer of Decatur, a local computer store, asked them to review a new commercial farm accounting software package, BOOKEYPER, written by David Fathauer, a local farmer. After reviewing this package, the Wentworths decided to use it in their operation.

Currently, they have 660 acres of land rented and own another 60 acres on which they produce corn and soybeans. Their major use of the microcomputer is for financial records and marketing accounts. Stratovisor and now Farm Dayta (Illinois Farm Bureau) give a market update every 10 minutes, and the computer is on every day from 9:30 A.M. to 1:15 P.M. Monday through Friday to capture this information from the FM radio transmission. They also use the computer for writing checks, farm accounting, assembling grain marketing plans, making operating decisions, and word processing. The Wentworths also own an Apple IIe with two disk drives and 64K, an NEC dot-matrix printer, and an 80-column monitor as well as an APPLEWRITER II word processing program and the VisiCalc electronic spreadsheet.

Steve and Pam Wentworth and their family are also using their microcomputer for educational purposes. They have software programs for spelling, reading, and mathematics. Pam is now teaching a high school computer course.

Thomas and Janet Fritz, who farm near Kankakee, Illinois, have been using their Commodore 4040 microcomputer since 1980. Their software and hardware choices were the result of working with a commercial software vendor who developed a farm accounting package specifically for them. The couple took two years to make their selections of hardware and software, and they considered the software service support of the vendor quite useful in getting started. Janet Fritz emphasizes the importance of reliable vendor service.

The Fritz family operates a farrow-to-finish hog operation and grow corn and soybeans on 190 acres of owned land and 345 acres of rented land. A farm accounting software package allows them to keep records similar to those maintained by the Farm Business Farm Management Service. With this program they prepare monthly projected cash flows, which they give to their banker. Actual cash flow reports are prepared at the end of each month to check the projections.

The Fritzes also use their microcomputer extensively for field records. For each field they keep a record of the fertilizer, chemicals, and seed that have been applied as well as a record of the fuel and repair costs. When crop yields are available, the profitability of the individual fields can be quickly determined by using a microcomputer farm records program and the DATA MANAGER program. Janet also keeps information on their hog enterprise in the microcomputer system. The cash flow planning reflects the costs and returns for their hog operation.

The major purposes for which the Fritz family use their microcomputer are farm accounting, cash flow planning, enterprise cost analysis, field-by-field enterprise accounting, and township assessor's accounting. Software packages purchased include WORD PRO 4, a word-processing program, and DATA MANAGER, a data base management system. These programs have helped the Fritzes maintain effective financial control of their business. In addition, they have accumulated about 40 educational programs for their family.

The planning, organizing, directing, coordinating, and controlling functions are more efficiently executed through the use of better farm record information, cash flow projections, and balance sheets. The total cost of the hardware and software for this operation was $3500.

John and Peggy Fechter live in White County, near Carmi, Illinois. The Fechters attended a number of University of Illinois Cooperative Extension Service meetings and visited computer stores prior to buying their IBM XT home computer. One of the first software programs John Fechter purchased was the LOTUS 1-2-3 spreadsheet.

Currently, their operation involves 1890 acres of rented and owned land. They operate as Fechter Farms, Inc., a corporation with nine different divisions under a Subchapter S corporation formed in 1984. Records are kept for nine different legal entities.

John has been maintaining farm records for six other landlords and tenants in a cash-grain operation. Each one of these partners requires separate accounts for income-tax reporting.

Prior to purchasing a microcomputer, the Fechters used to keep their farm records and accounts with the Illinois Farm

Business Farm Management Service. A microcomputer accounting package, however, was used concurrently with the hand records during the first year they used the microcomputer. Business reports and records kept on the microcomputer are financial records and accounts, information for the Internal Revenue Service 1040 F report, cash flow and budgeting information (actual and projected), loan activity summaries, family living expenses, enterprise profit and loss statements, payrolls, production by field, and inventories of grain. In addition, the microcomputer is used for writing checks.

The Fechters also subscribe to Professional Farmers Instant UPdate, a commercial marketing service that sends market information and advice, which is displayed on their computer screen. A linkage mechanism called a modem is necessary for this process.

Since purchasing the microcomputer, John and Peggy have attended two University of Illinois Cooperative Extension Service seminars at Southeastern Illinois College and two commercial microcomputer seminars.

Total accumulated cost of the Fechters' hardware and software was $12,800. The hardware now includes an IBM AT home computer with 256K, a 30-megabyte hard disk, an IBM color monitor, an Okidata 82A printer (for checks), an Okidata 83A printer (for forms), a switch box for the two printers, and a Hayes 300-baud Smartmodem. Their software includes FMS FINANCIAL MANAGER, FMS CROP MANAGER, LOTUS 1-2-3 spreadsheet, EASYWRITER, IBM Asynchronous Communications Support, and FRIENDLYWARE software packages. An Okidata 39.1 printer was purchased in the spring of 1989.

Because the FMS FINANCIAL MANAGER and the FMS CROP MANAGER software packages are interfaced, one entry is all that is needed. A check may be written or a deposit made and all accounts are automatically updated. For example, the bank balance can be increased or decreased, the grain bin inventory can be adjusted, and the loan balances are automatically brought up-to-date. Individual expenses, such as for machinery or animals, can also be tracked by the computer system.

Bill and Ruth Olthoff operate a 1,600-acre farm at Bourbonnais, Illinois. All the records were kept by hand in a double entry system until the beginning of 1984. The Olthoffs spent almost two years thinking and reading about microcomputers. Ruth enrolled in a number of business and accounting courses as well as a business computer course at Kankakee Community College. Then David Smith (a graduate of the University of Illinois Department of Agricultural Economics with a major in farm management and an employee of On-Farm Computing), a commercial vendor of hardware and software, demonstrated a farm records program to the Olthoffs. Smith left a microcomputer with them for a week to allow time for hands-on experience. This sent them on a serious search for the right software and hardware for their use. That's all it took; the Olthoffs now own a 256K IBM AT home computer with 50-megabyte hard disk storage.

During the first year, the Olthoffs decided to place their major emphasis on farm records. Their entire bookkeeping system was transferred to the computer immediately. The accounts-payable and check writing program operation were mastered during the first six-month period, and a payroll system was added. Particularly useful in their vegetable production enterprises is an effective payroll program for 75 different employees; the payroll program interfaces with the general ledger and accounts payable. LOTUS 1-2-3, WRITE word processing package, MITE communication package, and assorted templates also are being used.

Jake and Phyllis Salm are also clients of David Smith. For five years, Jake had maintained hand-kept records for the Coordinated Financial Statements developed by Dr. Tom Frey of the University of Illinois Agricultural Economics Department. The Salms own 300 acres and rent 910 acres in five separate tracts and need to keep separate landlord and tenant shares for each tract. The operation also includes a son who drives a farm-owned semitrailer truck during the off-season and jointly owns 200 acres. Separate accounting of expenses and receipts are necessary for these various ownership combinations. Smith recommended that the Salms use a commercial farm records program, GENERAL ON-FARM LEDGER. They also purchased a PEACHTREE electronic spreadsheet.

Starting in 1984, monthly farm records and accounts have been maintained and annual cash flows have been projected. Other financial statements that the Salms are maintaining on the computer are net worth, landlord and other-party settlements, loan activity summaries, and enterprise profit-and-loss statements. Crop expenses, including those for fertilizer and chemical applications, are recorded field by field.

The Salms are emphatic about the importance of the on-farm demonstration and training provided by their vendor. They say the vendor support service has been essential in helping the Salm family make effective use of the microcomputer during the first six-month period of ownership. In 1988, 30 megabytes of storage capacity were added to their IBM PC. The Salm's son and daughter-in-law, Michael and Janet Salm of St. Ann, Illinois, are now also heavy users of the microcomputer hardware and software.

Terry Lambert of Albion, Illinois, bought a 512K Radio Shack Tandy 3000 microcomputer in 1987. A copy of Tandy DESKMATE allows word processing, data base management, and spreadsheet activity. University of Illinois Cooperative Extension Service information and Farm Bureau meetings on microcomputers have permitted Terry to obtain the needed start-up information. The AG PAC Agricultural Software Package (including farm records, field records, beef break-even points, and marketing alternatives) was one of the initial software packages obtained. In addition, other more detailed farm record packages have been evaluated. Church records and Edward County Pork Producers Association records also are being kept, using DESKMATE integrated software. In the first two years Terry owned the microcomputer, approximately $3000 has been spent on both hardware and software. He farms about 913 acres of tillable land in corn, soybeans, wheat, cattle, hogs, and sheep.

Morris Bell, a farmer near Chandlerville, Illinois, started looking for a microcomputer in 1984. In April 1988 he purchased a 740K IBM PC compatible with a 20-megabyte hard disk, a Samsung color monitor, and an Epson Apex 80 printer. Software purchases included a LOTUS 1-2-3 Spreadsheet Release 2.01, a Farm Bureau Accounting package, and an IBM Writing Assistant word processor. Total cost was approximately $3200.

Morris has attended University of Illinois Agricultural Extension Service seminars. His microcomputer dealer also provided some hands-on experience prior to purchase. Morris has an interest in 370 acres of owned land and 645 acres of rented land. Corn, soybeans, wheat, snap beans, cucumbers, and popcorn have been raised.

Recently, Morris and David H. Crosnoe have collaborated in developing a software package, DBA Farmer's Fastline Software, copyright © 1989. This software package is designed to calculate variable costs and produce a farm field record system. The combined field operator's share of income and expenses and landlord's share of income and expenses can be obtained for one or more crops per year for each field and accumulated totals?

Ronald and Ann Christensen of Manlius, Illinois, made their purchase of a 512K IBM PC microcomputer 10-megabyte hard disk and Epson printer in 1983. Software purchased included the PC WRITE word processing package, TIPTERM, a communication package, and LOTUS 1-2-3 spreadsheet software.

These purchases followed an evaluation of hardware and software at the University of Illinois, Champaign-Urbana, by the Christensen's daughter and son. A commercial seminar also was attended in the Bureau County area.

Cash flow budgeting, farm records, livestock production records, and providing a communications connection for market information have been the major uses of the microcomputer through 1989. The cash flow budgeting is being jointly completed with the Farm Business Farm Management Service field representative for the 613 acres of corn and soybeans and

40 litters of hogs.

Dale and Christine Anderson, Bureau County, Tiskilwa, Illinois, started their evaluation of microcomputers in 1981. They studied microcomputer software and hardware together in the process of evaluation. A 128K Apple IIe microcomputer and an Epson FX80 printer were purchased in 1983. An AT & T PC compatible microcomputer with 30-megabyte capacity was purchased in 1986 for use with the Farm Management System farm record program. An EASYWORD word processing package was obtained along with the LOTUS 1-2-3 spreadsheet program. The FMS farm record system is being used to keep track of 632 tillable acres of corn, soybeans, and alfalfa and 200 litters of hogs. The Andersons also are enrolled in the Farm Business Farm Management Service.

Major uses have been in the farm business and hobby areas. A number of farm business decision aids are being used to develop cash flow estimates in completing financial planning. Also, market information has been combined with farm record cost analysis to produce more effective and efficient decision making. A number of programs for family use have been effective educational learning devices.

Eldon and Marilyn Eigsti of Buda, Illinois, began their two-year search for microcomputer hardware and software in 1985. They had attended a Bureau County Cooperative Extension Service meeting and the Farm Progress Show during the early stages of evaluation. The 640K Radio Shack Model 1000 with a 30-megabyte hard disk, a Sears Roebuck monitor, a 10X Star Gemini dot-matrix printer, and software were purchased in the spring of 1985 and later. Approximately $4200 had been invested in software and hardware by 1989.

Their software includes the TFS PROFESSIONAL WRITE word processing program, a SHARE WARE COMMUNICATION program, a DESKMATE spreadsheet, the FBS Systems TRANSACTION PLUS farm record program and the COORDINATED FINANCIAL STATEMENT program.

The *Personal Computing* computer magazine, a course at the local high school conducted by a junior college, and the

Farm Bureau have served as information sources. Eldon also provides the equipment service and information services for the Illinois Farm Bureau Farm Data System.

The Eigsti enterprise is a crop and beef-feeding farm operation with 464 acres of tillable land being farmed by Eldon.

The family's major microcomputer use has been with farm accounting, decision making aids, and camp and church records. Marilyn Eigsti is using the word processing package for correspondence.

Gerry and Sue Frank of Tampico, Illinois, began looking at microcomputer software and hardware in 1986. A 612K Epson Equity One Plus microcomputer, an Epson LX800 dot-matrix printer, a BETTER WORKING word processing program, a data base management program, a communication package, a graphics package, a spelling checker, an outliner program, and a general utilities package were purchased in 1988 for approximately $2200. Both software and hardware were evaluated together as the purchase was made.

Major uses have been for cash flow planning and government program alternative evaluation. A Great Plains Crop and Livestock Accounting program has been purchased. The learning process for this farm record program began in 1989. The Franks farm 614 acres. Gerry also has a nonfarm job with Empire tools in Sterling, Illinois, and Sue sells real estate and uses the microcomputer communications package to obtain current real estate listings.

SUMMARY: FARMERS USE A WIDE RANGE OF MICROCOMPUTER HARDWARE AND SOFTWARE

Microcomputer software application programs and packages for farmers and ranchers have been developed by commercial firms, universities, and public service institutions and more are being developed all the time. This microcomputer software provides farm or ranch managers with a means of determining how the various parts of their businesses or enterprises are performing. Production performance measures may be in yields per acre, milk per cow, pounds of beef per beef cow unit, pounds of pork produced per litter, or pounds of vegetables per acre, while financial performance may be measured in terms of net worth, financial ratios, or other financial and economic gauges.

The case studies of 12 farms indicate that a wide range of microcomputers and both hardware and software packages are being used. As lower-cost hardware and improved software packages were developed and additional needs became apparent, farmers have made additional purchases.

A tally of the 12 farms indicates that no two have the same identical hardware package. Nearly all of the farmers with farm accounting packages either purchased computers with sufficient RAM capacity and hard disk storage capacity or have added additional RAM and hard disks or hard cards to their computers for faster operation and additional storage capacity.

The initial interest among all of these farmers for buying a microcomputer was to aid in farm accounting and record keeping. The tally of the 12 farmers indicates that 9 have adopted a farm financial account record keeping program with numbers, weights, and values similar to those used by the Illinois Farm Business Farm Management Service. Only 4 added check writing enhancements to their programs. Seven of 9 have adopted various enterprise record keeping programs. Six out of 9 are using programs to project cash flow for total farm and/or farm enterprises. Nine of the farmers have included word processor packages in the set of software programs purchased.

A wider range of decision aid programs have been used. Many of these have been spreadsheet templates. In other cases, stand-alone programs have been used to evaluate alternate plans of action.

Gaining Access to Other Computers

Marketing decisions regarding crops and livestock are particularly crucial to the success of farmers and ranchers in the current agricultural scene. A number of farm operators are improving their marketing skills by securing other information sources than those traditionally used. Many commercial market information sources supply current data and projections via newsletters, reports, and electronic media.

One effective way of receiving daily market information is by linking a personal computer to a commercial information source by telephone. Any microcomputer owner can pay a monthly or an annual subscription fee to hook up to a computer at some distant location to obtain such timely information. The linkage between a telephone and a microcomputer is made with a hardware device called a **modem** (modulator/demodulator). This device plus a software communication program allows a person using a micro-computer to call an information source and request specific information. That information is then sent over the telephone line and appears on the computer display screen or monitor of the person requesting the information. The receiving computer can either print the information requested or store it on a disk for later display or printing.

One of the major marketing decisions is the pricing of commodities produced. Computer technology can be used for profitable pricing in three ways:

1. Market supply and demand information can be obtained via satellite dish and a dumb terminal (one that can only send and receive data) and from a time-share mainframe computer

(a large central computer) or minicomputer; via telephone, such as AgriData Network and Professional Farmers Instant Update; via radio or television transmission; or by a combination of these systems. A microcomputer can be used as a terminal to access remote computer-based information systems by telephone.

2. Analysis of the market supply and demand situations can be obtained by reading newsletters, listening to radio and telephone reports, and accessing computer market information sources.

3. Microcomputers can be used to store commodity prices and provide moving average price information for plotting and charting 3-day, 5-day, and/or 15-day periods.

For profitable pricing, primary decision makers for farm and ranch operations must obtain adequate information by using one or more of these three sources. Improved marketing and pricing decisions depend on the information available.

A number of commercial market information sources, such as AgriData Network, Commodity Communication Corporation, and Professional Farmers Instant Update, provide data for pricing commodities produced. Agricultural markets, financial markets, weather information, and transportation and barge information are all available, as are agricultural market analyses and suggested marketing strategies. As with most commercial systems, users are charged an annual fee plus monthly user costs.

Newsletters are also provided by commercial firms. The newsletters are usually available in a time period of three to five days from the release of information. AgriData Resources, Inc., supplier of AgriData Network, for example, publishes *Farm Futures* magazine and a weekly newsletter called *Top Farmer Intelligence.* Microcomputers with a modem can supply the same information within minutes or hours. In some cases, receiving market information immediately can help in making timely decisions.

Following are public and commercial sources of market information that have been available in the past few years:

Name	Address	Information Supplied
American Agricultural Communications (AAC) System	American Farm Bureau Federation Agricultural Communications System 225 Touhy Park Ridge, IL 60068	Retrieval, market, weather, legislative, analyses, and advice
AgLine	Doane Publishing 11701 Borman Drive St. Louis, MO 63146	Nationwide electronic mail, agricultural software reviews, public domain agricultural software, market advice, and analyses
AGRICOLA	USDA National Agricultural Library, Information Systems Division 10301 Baltimore Blvd. Beltsville, MD 20705	National service for farmers and agricultural researchers interested in agricultural research and extension publications
AgriData Network	AgriData Resources 330 E. Kilbourn Milwaukee, WI 53202	Nationwide source of agricultural information on cash and futures prices, commodity analysis, financial markets, and transportation; updated marketing analysis and advice
ANSER	Agricultural Data Center University of Kentucky Lexington, KY 40546	Problem solving, information retrieval for management information systems (primary users are extension staff in Kentucky)
Commodity Information, Inc.	Commodity Information, Inc. 560 South State St., Suite E-1 Orem, Utah 84058	Provides commodity market information along with color charts and graphs of each plus futures prices, USDA reports, market analyses, and hedging records; financial and production management records also available

CompuServe	Information Service Division 5000 Arlington Center Blvd. P.O. Box 20212 Columbus, OH 43220	Provides personal computer owners with public access to online communication and information services
Computrac, Inc.	Technical Analysis Group P.O. Box 15951 New Orleans, LA 70175-5951	Charting and analysis software
CRIS	Current Research Information System 5th Floor, NAL Building Beltsville, MD 20705	Nationwide information retrieval for publicly funded agricultural and forestry research
Martin Marietta Data System	Martin Marietta Suite 640 2340 S. Arlington Heights Rd. Arlington Heights, IL 60005	USDA reports on a monthly or an hourly fee basis
National Weather	Climate Analysis Center World Weather Bldg. Room 201 Washington, DC 20233	Nationwide temperature, precipitation, and various weather reports
Professional Farmers Instant Update, GlobaLink	Professional Farmers of America, Inc. Instant Update 219 Parkade Cedar Falls, IA 50613	Nationwide information retrieval of commodity prices, market advice, analyses, and charts

All these networks provide similar services. Write to the individual companies that interest you to learn about the specific details of service supplied. Information on these programs can also be found in the Agricultural Computing Source Sheet (revised), 84 Communications, Doane Publishing, 11701 Borman Drive, St. Louis, MO 63146.

SUMMARY: USING THESE APPROACHES FOR PROFITABLE PRICING DECISIONS

Because sound information is needed on the latest supply and demand situation by commodity in order to make good decisions, a specific information source should be selected for your farm and agribusiness needs. In addition, a strategy for pricing and hedging information must be selected from commodity newsletters, computerized sources, commodity newswire services, and other sources.

If a computer terminal or a microcomputer used as a terminal is available, computer-based information systems can be accessed when you need information on a particular commodity. Microcomputer software plus a modem allows access to a large mainframe computer system, a minicomputer, or another microcomputer.

Microcomputer software has been developed to provide complete market charting analysis by commodity. For example, one Illinois farmer has been obtaining price information daily from a commercial information source. He can enter the commodity price data on his computer using the software plotting program for 3-day, 5-day and 15-day moving averages. During 1981–1982, he sold his crop at prices considerably higher than his county averages. He also could have subscribed to a service from a commercial firm that would have provided the same information by printing it out using a modem and his printer. As a general rule, pricing and market management possibilities are understood and used more effectively when reliable supply and demand information, charts, and pricing information and strategies are available. The task of obtaining and organizing details is done faster and better by computer.

Farm business records also are essential in determining the production costs of commodities. Farmers with microcomputer software packages for their farm accounting and record keeping that allow enterprise cost accounting have vital data at their fingertips for making profitable pricing decisions. Of course, goals and marketing strategies are also essential for profitable pricing.

In summary, three approaches are open to farmers choosing a market information system:

1. Purchase a commercial source of agricultural market information with a complete package of computer software and hardware.

2. Secure a combination of commercial and public sources of market information. A computer terminal, microcomputer, satellite disk, FM radio sideband transmission with microcomputer software, and other types of equipment can be purchased to access particular information systems.

3. Purchase a microcomputer to perform bookkeeping, inventory control, or information retrieval and planning functions. Part of this approach may be a combination of commercial and university newsletters, commercial and public computer information systems, and individually constructed microcomputer software to develop decision making aids.

Microcomputer Software

Software refers to the information that makes a computer operate and perform some function. It can be written down or encoded on some other recording medium but is not otherwise a tangible, physical object. The best-known example of software is a **program,** or the list of very detailed coded instructions that tell the computer how to carry out the steps of some information processing procedure. Without software, any computer is a completely useless assemblage of mechanical and electronic junk.

We may say that almost all software is of four main types: programming language translators, applications programs and packages, operating systems, and utility programs.

PROGRAMMING LANGUAGE TRANSLATORS

A language **translator** is actually a special-purpose program whose function is to decode a program written in some standard computer programming language and translate it into **binary** form—the only kind of "language" the computer can ultimately understand and respond to.

Programming language translators are generally of three types: interpreters, compilers, and assemblers.

Interpreters

An **interpreter** attempts to decode and execute (run) a program line by line. This requires little memory space to implement and makes for flexibility, but it is inevitably a relatively slow and inefficient process. The BASIC (Beginner's All-purpose Symbolic Instruction Code) language is usually implemented as an interpreter in most microcomputers. This is perhaps the easiest computer language for most people to learn and has been by far the most widely used on microcomputers. Here is an example of a very simple program in BASIC:

```
10 DATA 3, 5
20 READ A, B
30 LET C = A + B
40 PRINT C
50 END
```

This program assigns the constant values 3 and 5 to the variables A and B respectively, adds them to get C, displays the result of the operation on the computer's monitor screen, then stops. (See Fig. 5.1 for another BASIC program.)

Compilers

BASIC can also be implemented as a **compiler** or translator that decodes an entire program from top to bottom; the resulting binary output is then loaded into the computer's memory and executed (run). This approach is far more efficient than line-by-line interpretation, when the program is correct, but it requires much more memory, and the program must be repeatedly recompiled until there are no more errors or "bugs" in it.

The next most popular language for programming microcomputers is probably **Pascal**, which is favored by computer science professionals because it encourages the formation of good programming habits and the writing of clearly correct and "structured" programs in a way that BASIC does not. Other programming languages sometimes implemented as compilers include the very concise and efficient **C** and **FORTH**; a graphics-oriented language suitable for children called **LOGO**; and the two old standards on larger computers, **Fortran** and **COBOL**.

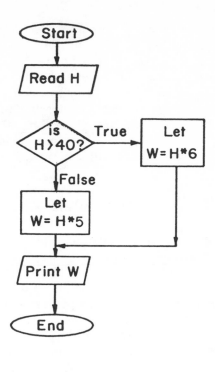

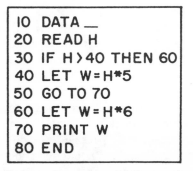

```
10  DATA __
20  READ H
30  IF H>40 THEN 60
40  LET W=H*5
50  GO TO 70
60  LET W=H*6
70  PRINT W
80  END
```

Fig. 5.1. Simple program in BASIC: An employee is paid $5 per hour for working 40 hours or less during a week and $6 per hour for working more than 40 hours. How much should the employee be paid for working H hours?

Assemblers

An **assembler** allows use of a "lower-level" language with a close relationship to the physical structure or hardware organization of the computer. Assembly language is thus highly efficient and flexible but correspondingly complex, technical, and difficult to learn. It is therefore used for the most part only by expert programmers. Another disadvantage is that it is very specific to particular models of computers (CPUs), so an assembler program written for one brand of computer ordinarily cannot be used with any other brand.

APPLICATIONS PROGRAMS AND PACKAGES

An **applications package** is a program or interrelated group of programs that accomplishes a set of related tasks such as accounting, preparation of printed matter ("word processing"), or sending information to other computers. The package is normally offered and purchased as a unit. Some of the principal types of preprogrammed packages are those used for accounting, spreadsheet analysis, database management, word processing, graphics, and communications:

1. For accounting and financial and tax reporting, many packages are available that carry out all the usual accounting functions: general ledger, accounts payable, accounts receivable, checkbook balancing, and payroll (if needed). Others can prepare and print financial statements such as the balance sheet, income statement, funds statement, and cash flow projections.

2. Especially useful for financial planning, forecasting, and analysis of the hypothetical or "what if" type of situation is the electronic spreadsheet package in which the computer screen becomes a "window" that can be moved around anywhere over a large worksheet laid out as a grid of rows and columns (similar to an accountant's ledger sheet) (see Fig. 5.2). Cells of the grid can contain labels (words), numbers, or formulas that make use of numbers in other cells to calculate and display numeric results such as totals and averages. When

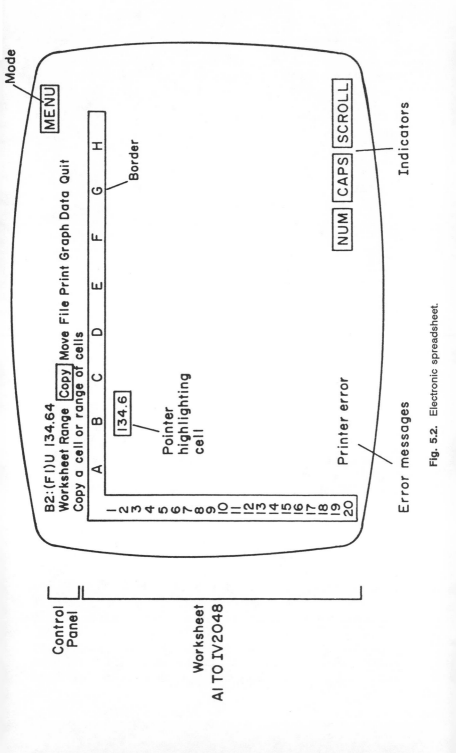

Fig. 5.2. Electronic spreadsheet.

a number in any cell is changed, all the others that depend on it are automatically and quickly updated to reflect the change.

Examples are LOTUS 1-2-3, Microsoft Excel, Multiplan.

3. Data base packages are generally of the record keeping on inventory control type and are used for recording, locating, and summarizing such information as production or breeding records, etc., then printing out reports based on that information.

Examples are dBASE, R:base, Paradox.

4. A word processing package is a specialized program for creating and editing textual (written) material, formatting it neatly, and producing attractive printouts of any sort of document.

Examples are WordPerfect, Microsoft Word, MultiMate.

5. Graphics packages allow the production of graphs, charts, maps, diagrams, signs, animation, and other pictures and images when used in conjunction with appropriate hardware such as plotters, color monitors, and special printers. Complex information can often be summarized in a much more understandable way by using a graphic representation: as the saying goes, "One picture is worth 1000 words."

Examples are AutoCAD, MacDraw, Cricket Graph.

6. Communications packages allow your computer to communicate with other computers over a telephone line. Your microcomputer can function merely as a "dumb" terminal in accessing another computer system, or it may have the "smarts" to transfer entire files of information to and from the remote computer (called uploading and downloading).

Examples are Kermit, Crosstalk, Smartcom.

7. A recent trend has been the development of **integrated packages** combining several functions, usually at least word processing, spreadsheet, database, communications, and simple graphics. The major advantage is the assurance of compatibility between all functions; for example, a portion of a spreadsheet can easily be brought over and inserted into a word processing document and vice versa. A disadvantage is that the components may not all be of high quality, and the purchaser may have to pay for things he or she doesn't need.

Examples are Framework, Symphony, Reflex.

8. Other programs and packages are described below:

Project management programs are available that use sophisticated quantitative tools such as Critical Path Method (CPM) and Program Evaluation and Review Technique (PERT) to keep track of ongoing activities and projects, keeping the user continuously up-to-date about progress, deadlines, milestones, bottlenecks, priorities, and other crucial operational factors.

Examples are Harvard Total Product Manager, ViewPoint, Timeline.

Many programs and packages are available for instruction in all subjects and other areas of knowledge. Some provide tutoring in specific abilities (such as how to use a certain computer package) or training in widely useful skills (such as typewriting).

Finally, many video game packages are available for home entertainment, ranging from extremely simple games for young children to highly elaborate products designed to absorb adults for many hours.

In addition to these standard and widely used general-purpose packages, many programs exist for use in specialized business and professional areas (for example, in the analysis of investment and production operations decisions), often using sophisticated decision criteria such as net present value or internal rate of return. Certain other packages permit the analysis and modeling of complex relationships such as those that calculate an optimal mix of fertilizers or least-cost rations for livestock. Others simulate market movements and provide a useful forecasting tool.

OPERATING SYSTEMS

An **operating system** is a special software package that controls the running of other programs and the manipulation of files. It allocates the computer's hardware resources to various system functions such as creating, copying, executing (running), and deleting files. Because most microcomputers use disks for recording these files, the operating system is known as a disk operating system (DOS, pronounced "doss"). An operating system is sometimes also known as the monitor, supervisor, or master control program.

An example of using a DOS is the operation of formatting or initializing a new diskette. The basic pattern of tracks and sectors into which information later written on the diskette will be organized must be established before the diskette can be used for anything further.

Some disk operating systems in common use include the following:

1. There is the "native" DOS that comes with most microcomputers when they are purchased. Many users eventually find these to be primitive and limiting, so software developers offer a number of more sophisticated and flexible DOS packages such as CP/M.

2. The CP/M (Control Program/Microcomputer) family of operating systems was for several years dominant on most of the older and smaller microcomputers and in many ways became a standard for program and package developers. Having it makes available to the user a vast range of software that was written for this DOS. Newer and more advanced versions are available, but CP/M seems to be losing ground to DOS packages written specifically for more recent computers.

One of these is the MS-DOS operating system used (under the name PC-DOS) on the IBM Personal Computer and the many PC-compatibles (a growing number of which are built in Japan).

3. The UNIX family, in its many variants and spin-offs, is descended from a highly sophisticated operating system developed for minicomputers. Some say that it, or something like it, will be the most popular future DOS. But others point out that, while programming experts like UNIX, it is not very **user-friendly** or suitable for beginners.

4. Another DOS is the UCSD Pascal p-System, which has the advantage of making programs and files portable from one make and model of computer to another.

5. Finally, a few prefer the Pick operating system, which has a number of variants under various names. The Pick DOS

is to this point not as widely used as some of the above, but those who use it seem enthusiastic.

UTILITY PROGRAMS

Often important system **utilities** such as programs for editing, sorting, debugging, and loading other files are included with the operating system. Sometimes they are sold as separate packages. Some are very useful, such as good general-purpose text editor and the utility programs that help recover damaged and deleted files.

GUIDELINES FOR EVALUATING SOFTWARE

Here are some considerations and questions to think through and to which you should obtain adequate answers before committing yourself to any specific software.

Does it do what you want done?

In thinking about whether to computerize, it is essential to do an informal "system analysis" of the present and probable future information needs of your operation. Where does the information that you need come from? Is information being acquired and stored that you could get along without, since it is no longer really needed? What reports (financial statements, tax returns, etc.) are required by those outside your operation, and what kind could be used to improve internal managerial decision making? What additional information could contribute to the success of your operation if only you could acquire, process, and use it in a fast and efficient manner?

User-friendliness

Does use of the program correspond to most people's normal and natural ways of thinking, or does it require you to be a programmer or computer expert in order to understand it? Is it easy to learn and easy to use? (Note that these are somewhat separate, and sometimes conflicting, considerations.)

Error handling and data editing

Does the program "hang up" or "drop dead" when wrong data are entered or the user hits a wrong key? Ideally, it should be robust enough to survive any kind of ignorant, random, or malicious input ("bulletproof"). It should be tolerant or forgiving of minor input errors, with no danger of losing valuable files and data if some unforeseeable accident occurs (such as a disk filling up unexpectedly).

Documentation

The accompanying descriptive materials (directions, manuals, tutorials, etc.) should be clear and complete, with ample helps provided both within the program and in comprehensive printed documentation.

A careful evaluation of software documentation means a thorough examination of the information provided in the overview. The overview should include a general description of the software package that provides an explanation of what the program can do and information about what functions it can perform. The overview should specify the particular microcomputer and the specific operating system software that is compatible. The computer language used, the capacity required, and the printing options should also be discussed.

Another part of software documentation to study is the tutorial section. Instructions demonstrate how to use the software program. Commands, procedures, illustrations, and other general discussion should give the potential microcomputer owner an idea of how the software program operates and functions.

A summary of the command sections in the documentation should be reviewed. Descriptions of commands should be clearly written. Command examples are usually included to aid understanding, and sometimes cards or templates are provided that can be placed on the keyboard for easier use. Specific tutorial diskettes also are provided with some microcomputers.

Essentially, the tutorial diskette teaches the operator how to work with a particular set of commands.

Another thing to look for when evaluating software documentation is the troubleshooting section, which should provide information on what to do if error messages appear on the screen while the program is in operation. A description of the steps to take if such messages are encountered should be included in the documentation materials or in the user manual.

The technical section should also be part of the documentation, including the blueprint of the microcomputer program or the flowchart. Information should be available on whether modifications can be made to the original software program. (Be aware that some vendors will not support their product if there has been software computer program modification.)

Finally, the documentation should include a table of contents and a detailed index, which save a great deal of time when you look for solutions to problems encountered while using the software.

Vendor Support

Does the program's distributor provide bug fixes, updated and improved versions, etc., free or at low cost? Is there a toll-free technical "hotline" number you can call with questions? (If there is, can you actually get through to someone knowledgeable and get an answer in a reasonable length of time?)

Some vendors are stronger suppliers of services than others. Helpful services to inquire about include (1) a telephone hotline, (2) warranties, (3) availability of low-cost backup copies and updates, (4) on-site training, (5) dealer repair and maintenance, and (6) vendor-sponsored newsletters and seminars.

The telephone hotline is a service that many vendors supply, although they encourage microcomputer owners to read the user manuals before using the hotline. A short telephone conversation, however, is sometimes the only way to clear up a question, so this service is an important one to look for in choosing a software vendor.

A written warranty for a software program should be available, one stating that the vendor will fix a program if it does not run. The exact operating-system software and hardware components required by the software should be stated in the warranty. The rights of the software purchaser should also be specifically described in the warranty.

Sometimes original diskettes fail to operate correctly, so low-cost backup copies should be available for any software program. The availability of such replacement copies of microcomputer software programs and the willingness of a vendor to supply them is another important factor to consider when choosing software. Over the years, vendors often make changes in the microcomputer software programs, so it is also important to find out whether a vendor will supply low-cost copies of updated software programs. Sometimes such updates are available only for one year.

Another service to inquire about from both software and hardware vendors is the availability of on-site (on-farm) training. Some vendors insist that on-farm training sessions be combined with a seminar or training sessions from their firm. In such cases, the cost of the training and seminars may be included in the purchase price of the hardware and software.

Although the availability of good dealer repair and maintenance service may be associated with training sessions for new microcomputer hardware and software owners, it is usually a good idea to discuss the amount of time that can elapse between the time service is requested and when the service is performed.

A final factor to look for in vendor service is the availability of newsletters and seminars. Newsletters are one way publishers or vendors of microcomputer software keep customers informed of new programs and changes in old ones. Vendors may also use newsletters to announce upcoming seminars. Such seminars allow you to see how other users are implementing a particular software package in their own operations.

Program Quality

Another part of evaluation is determining how well a particular software program performs. The following are specific features to look for when evaluating software performance: (1) adequate menus (see below) and prompts (request from the computer for further information), (2) simple handling of errors, (3) programming accuracy, (4) operating speed, (5) options for printing summaries or reports, (6) ease of use, (7) ease of modification, and (8) agricultural usefulness. Each one of these features will be discussed below.

A **menu,** in computer terminology, is a listing on the screen of available functions from which the user can choose. A software applications program or package that displays a menu on the screen or monitor is usually easy to use. The evaluation of a software program should involve some hands-on experience that allows you to see how menus are used. Options are usually listed on the screen by number. When you select a number and depress the RETURN or ENTER key, that option will be put onto the screen to help you perform a particular operation. Menus increase the speed of operation of the microcomputer software applications program.

Another factor to look for in evaluation of the application of software is how errors are handled. Hands-on experience is essential to see what happens if a wrong entry is made; usually, instructions will appear on the screen. The speed of correcting an entry is of prime consideration. Seek software programs that handle errors with speed and accuracy.

Program accuracy is essential for successfully using and operating software. One evaluation technique is to make sure the data that is entered is processed correctly. To do this, use sample data and check results when the end summary is known; for example, enter farm record summaries and the previous year's data to check a farm records software program.

Speed of operation is also essential for efficient use of software. Hands-on experience provides an opportunity to determine the operating speed of a particular software program or package. One person, or a team of two, should be allocated the time necessary to perform this essential step. It may take a number of days or weeks to enter data for specific bookkeeping, inventory control, and decision-making programs to determine the speed of entry, computation, and final printing of reports of data summaries. Specific software programs should be tested with data similar to that used in your own farm operation.

Printout options also should be checked for usability. Hands-on experience allows you to see the printed text from different segments of a software program. These printed reports and summaries are valuable decision-making aids and also provide records for other purposes. For example, a copy of a projected cash flow is useful when arrangements are made for obtaining operating funds from a bank. Projected and actual cash flows are also useful in maintaining credit lines with most financial institutions, such as banks, Production Credit Associations, the Farmers Home Administration, and private individuals. If you cannot generate a printed summary with your own data when trying out a software program, most firms will supply detailed examples of the reports and summary information their programs can generate.

A software program or package requires thorough checking and testing. Hands-on experience over a period of days allows the potential microcomputer owner and/or operator to judge how easy the program is to use. Menus, instruction when errors are encountered, how fast the computations and printed materials are presented, and the clarity of instructions are all factors in the ease of use.

Ease of modification is another evaluation factor. Software vendors do not all have the same rules on modifying original programs. Generally, if the vendor includes support of the software program or package in the original purchase price, the original program must be maintained.

Vendors who do not support changes in the written computer program will make allowances for program changes. A listing or printed statements of the actual computer program allow an individual to modify it. It may be necessary to modify a program to provide information unique to a particular farm or ranch operation.

Evaluation of software applications programs for farm

operations should always include an agricultural usefulness review. Each farm or ranch has particular information needs. For example, you may need certain information for daily management and decisions, while information on expenses, receipts, inventories, tax data, labor data, and enterprise performance data is needed only on a monthly or annual basis.

Evaluation of an applications package should involve how well daily, monthly, and annual information is provided to key decision makers. You must make a judgment based on review, hands-on experience, and study of reports and results of a particular software program. Checklists are available to evaluate farm record and accounting software, but you may want to list particular characteristics that you desire as well. Allocate sufficient time for this phase of evaluation.

In the long run, an agriculturally useful microcomputer software applications program with menus and prompts—one that handles errors with ease, is accurate, operates with speed, provides adequate printout options, is easy to use, and can be modified—will increase the efficiency of a farm or ranch operation.

SOURCES OF INFORMATION

How do you find out more about what software is available and whether it is any good?

1. Look in computer magazines for ads and reviews. Advertising claims cannot always be taken at face value, of course. Most computer magazines publish reviews in every issue and periodic comparative reviews of several packages of a given type (light-duty word processors, graphics packages, etc.).

2. Talk to other users. You can locate others with similar interests and problems by joining (or starting) a computer club or by taking a computer course offered by the local school, college, or computer store.

3. Find a consultant (who is *not* a salesperson!). A good consultant will be worth every cent of the $50 to $100 an hour you will probably have to pay to steer you to the best package

for your type of application. Demand a guarantee that the consultant has no business relationship with any vendor of hardware or software, since any such interest is likely to distort judgments and advice about what is best for you.

WAYS OF OBTAINING SOFTWARE

Here are some ways of obtaining useful software:

1. Buy it off the shelf or by mail. Most computer stores stock a number of the better-known programs, which may be purchased on diskettes. Some may be willing to special-order software for you. Computer magazines are full of advertising by mail-order discount software houses, but the buyer must beware of slow service (or no service), no local support or follow-up, and, in short, all the usual problems of buying anything by mail.

2. Write it yourself. *Be warned* that many have tried this, usually resulting in months of frustration and, almost invariably, in having to buy software anyway. *But* you may be the one person in a thousand who actually has the time, patience, and talent necessary to do this.

3. Hire a consultant or professional programmer to custom write programs. This may seem expensive, but could well turn out to be cost effective in saving you time and headaches and in tailoring the software to the special, unique needs of your operation or management practices.

4. Acquire it by theft, by making unauthorized copies of copyrighted software. Many software developers try to prevent this by making their programs noncopyable, but this is only partially successful, and much illicit software is in circulation for which the developer has not received fair compensation. Copying, using, and giving (or, worse yet, selling) such software to others is illegal, unethical, and sleazy.

5. Copy the many program listings published in books and magazines. This normally does not raise a legal or ethical problem if the program is exclusively for your own use and is not to be sold. However, this practice requires a sizable

investment of someone's time in entering and debugging the program, and the quality of the result may be very poor.

6. Take advantage of low-cost shareware and free public-domain programs. Shareware is a term for utility and applications packages that are freely copyable but for which the developer requests a donation of some $5 to $100 from those who find the program useful and would like to obtain more documentation, bug fixes, updated versions, and other continuing support. Many free public-domain programs are available from individuals, bulletin board services, user clubs, colleges, etc. This software is completely unsupported, of course, and much of it is of inferior quality, but a great deal of surprisingly good and useful software is being given away and can be acquired almost for the asking.

A FINAL CAUTION

Remember to budget *at least* as much for software as for hardware! Stinginess about purchasing software is usually a false economy and a major reason why a disturbingly large percentage of small computers are sitting around unused and gathering dust a few months after purchase.

REFERENCE

Gear, C. William. *Computer Organization and Programming with an Emphasis on the Personal Computer,* 4th ed. New York: McGraw-Hill, 1985.

6 Microcomputer Hardware

By **hardware** we mean the computer itself or all of the physical, tangible parts of a computer system: the keyboard, disk drives, monitor screen, connecting cables, etc. Everything that you can see and touch, in other words, every mechanical and electronic device that enables a computer system to operate, is referred to as hardware.

The principal functional units of any computer may be represented as shown in Figure 6.1.

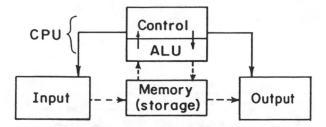

Fig. 6.1. Basic function units of any computer, showing status and control information (–) and data and instruction function (----), the central processing unit (CPU), and arithmetic-logical unit (ALU).

INPUT DEVICES

Input devices read an input medium and convert the information recorded there into the only thing the computer can actually process–patterns of electronic pulses. See Figure 6.2.

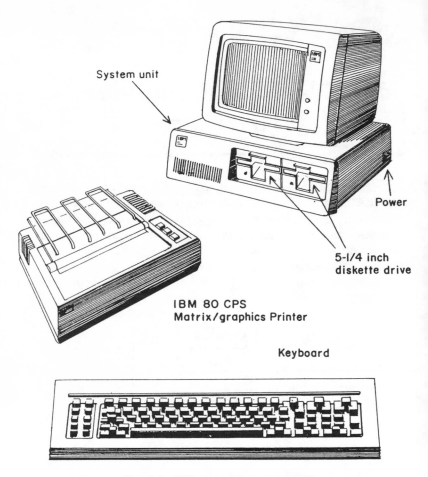

System unit

Power

5-1/4 inch
diskette drive

IBM 80 CPS
Matrix/graphics Printer

Keyboard

Fig. 6.2. IBM personal computer (PC).

The **keyboard,** for example, generates a different pulse/no pulse pattern or sequence for each key pressed; this sequence is then sent to the computer for processing (after we press ENTER or RETURN). We may think of the information as consisting of patterns of the **binary digits (bits)** 1 and 0, corresponding to "pulse" and "no pulse." Information is normally handled in 8-bit sequences called **bytes**. Many of the newer machines can process two or four bytes at a time if they have 16- or 32-bit **microprocessors** or "chips."

Computers can also "read" information encoded as patterns of tiny magnetized spots on a recording medium such as a tape or disk. Files containing data or programs to be input to the computer can thus be created on a magnetic tape or disk (usually by another computer), then stored and input to the computer's memory (internal storage) as needed.

Information has sometimes been recorded in the tracks of ordinary audio cassette tape, then input to a microcomputer using a compact cassette player of the kind commonly sold for $25 to $40. But information transfer from cassette tapes is notoriously slow and unreliable and generally to be avoided by serious microcomputer users. Diskette drives operate far more reliably and at least 20 times faster than tape cassettes.

Input information is organized on computer disks into concentric circular tracks and pie-shaped sectors. The recording medium is a plastic disk coated on both sides with iron oxide, which can retain the tiny patterns of magnetized dots in which the information is recorded. The two most common types currently are:

1. Flexible ("floppy") disks about 5.25 inches in diameter, often called "minifloppy" diskettes. These come in two **densities** of recorded information: "double density," holding about 360,000 bytes of information (about 180 typewritten pages); and "high density" in the PC AT, holding about 1.2 million bytes (almost four times as much).

Unfortunately, the older, widely used 5.25-inch diskettes varied greatly in their specifications and formats: some were single-sided, others double-sided; some were single density, others double density; still others were of higher density; some were soft-sectored, others hard-sectored; and they were formatted in varying numbers of tracks and **sectors**. It is rarely the case that a diskette created (formatted and written) by one brand of computer can be read by any other brand using a different operating system. An exception is the growing number of "compatibles" or "workalikes" designed to duplicate the most popular brands as closely as legally possible.

2.   The 3.5-inch "hardshell" diskettes (used in, for example, the Apple Macintosh and IBM PS/2 series and compatibles). These also exist in two densities: double density, holding about 720,000 to 800,000 bytes, and high density, holding over a million bytes each.

Applications calling for faster access to data or for very large data files require a "hard disk," which is a rigid metal disk spinning at high speed. A typical hard disk unit will store 5 million to 100 or more million bytes (1 million bytes = 1 megabyte) of information. A number of small hard-disk units are now available and have become feasible options for the users of larger data files.

Many other types of input devices exist but are probably of little interest to most individual and small business users. These include magnetic ink character recognition (MICR), used on the preprinted checks obtained through most banks; optical character recognition (OCR), used in the "bar codes" now seen on most grocery items in supermarkets; and a variety of analog-to-digital (A/D) converters that transform some physical measurement or motion (such as pressing a joystick or moving a mouse) into a digital input signal.

THE MEMORY

The computer can process and make use of only the information that is currently contained (represented electronically) in its memory or storage unit. Part of this memory is permanent; part is temporary.

The permanent part is called **read-only memory** (ROM), which, as its name indicates, can be read from but not written to (altered) by action of the computer. ROM usually contains utility software encoded in it that will be needed frequently and permanently by the computer, such as a "bootstrap" routine or the ability to read in an operating system file from a diskette. Sometimes a programming language translator such as a BASIC interpreter is also included in the ROM. (See

Chapter 5.)

The temporary part is called **random-access memory** (RAM), where information can be read from or written to anytime. The information in RAM is volatile, meaning that it disappears when the computer is turned off, while that in ROM is always available whenever the computer is turned on.

The size of the computer's memory (the volume of information it can hold) is usually given in multiples of 1024 bytes, or one kilobyte (KB or K). Thus a computer with 64 kilobytes of RAM would have $64 \times 1024 = 65,536$ user-accessible memory cells available for holding items of information.

The larger the memory, the more varied, complex, and valuable are the applications to which the computer can be put. Small memory capacity severely limits the usefulness of the computer and is in fact the principal constraint (other than lack of appropriate software) that prevents microcomputers from realizing their full potential usefulness in most situations.

Memory is often expandable by 64K plug-in units costing some $50 to $150 each, up to a designed-in limit for each model of computer.

THE CENTRAL PROCESSING UNIT (CPU)

The CPU is mostly contained on the **microprocessor** or "chip" that is the computer's "brain." It consists of two main subunits:

1. The **control unit** coordinates the operations of all other units of the computer. This component of the CPU contains the system clock, which emits electronic pulses through the computer's circuits at rates of some 4 to 30 million cycles per second, or 4 to 30 megahertz (MHz). These pulses are used to synchronize all other internal operations of the computer. Other things being equal, a higher clock rate means faster processing, up to a point where reliability becomes a problem.

2. The **arithmetic-logical unit** (ALU) contains the registers where information is represented electronically while

mathematical and logical operations are carried out on it. Registers have various widths, depending on the number of binary "bits" of information they can hold and transform at one time. Other things being equal, a 32-bit CPU should be about twice as powerful as a 16-bit CPU; in other words, it should be able to process twice as much information per second. A wider ALU also allows the computer to store and locate items in a larger "address space," permitting a larger potential memory size.

Here are some examples of widely used microprocessors:

1. Intel 8088: used in the IBM PC and many compatibles. (This is actually a "hybrid" chip with 16-bit registers but only an 8-bit data path, or bus.)

2. Intel 8086: a true 16-bit chip with more power than the 8088.

3. Intel 80286: a hybrid 16/24-bit chip used in the IBM PC AT and a number of compatibles.

4. Motorola MC 68000: a hybrid 16/32 bit chip used in the Apple Macintosh, Commodore Amiga, and Atari 520 st.

5. Intel 80386: a 32-bit chip now used in a number of IBM compatibles.

6. Intel i486: a new high-capacity chip that will become available in the early 1990s.

The two dominant types of microprocessors are thus the Intel 80x86 series used in the IBM PC–PS/2 and compatibles and the Motorola 68x0 series used in the Apple Macintosh, which is rapidly beginning to challenge IBM's dominance of the microcomputer field.

OUTPUT DEVICES

Output devices receive patterns of bits from the computer and convert them into something that can be understood by a human user, such as a character on a TV-like screen. This **monitor** screen is sometimes referred to as a CRT (**cathode ray tube**), after its principal component. It **echoes** the input typed at the keyboard, with previous lines of information

scrolling upward off the screen as new lines of information are typed at the bottom. The results of running programs are also displayed here.

Use of the computer for business or professional purposes normally requires a good-quality **printer** for producing hard copy reports, statements, records of results, and other documents. Printers are generally of two types: impact and nonimpact.

Impact printers operate on the principle of striking the paper through an inked ribbon, much like an ordinary typewriter. The more common (and less expensive) type of impact printer is the **dot-matrix** printer, in which each character is formed at the moment of impact by electronically activating a certain pattern of tiny rods within a grid arrangement. These rods strike through the ribbon, blackening the paper with that pattern of ink dots; then the print head moves on to the next position. A serviceable dot-matrix printer can be purchased for as little as $200, although the more rugged and higher-speed models with such extra features as graphics capability can easily cost over $1000 more. A dot-matrix printer with a fresh ribbon can produce quite acceptable reports for internal use. Some of the newer models are claimed to be able to produce near letter quality copy and thus to be acceptable in certain word processing applications in place of a typewriter.

However, correspondence-quality printing is consistently achieved only by using a **formed-character printer** of the "daisy wheel" or "thimble" type, which works by spinning a character "petal" into place (like a single-element or "ball" typewriter), then striking it from behind with a tiny hammer. Daisy-wheel printers are both slower and much more expensive than dot-matrix printers but have long been the only choice for those who require reliably high-quality printouts similar in appearance to those produced by an office typewriter.

The high-volume "line printers" used at large computer installations are usually impact printers of the print-chain or print-wheel type, producing hundreds of lines of output per minute. We are assuming here that the average individual or small business has no need for such high-capacity equipment. Most will find the typical dot-matrix printer, working at about 80 to 100 characters per second (cps), or a daisy-wheel model at some 40 cps wholly adequate for their operations. Some will, in addition, want to consider whether further graphics capability (plotting graphs, making charts) may also be worthwhile for them.

Three types of **nonimpact** printers may be mentioned: **thermal, ink-jet,** and **laser.** Many thermal printers are inexpensive but produce low-quality printing on special chemically coated paper that is difficult to handle and store. Ink-jet printers are far faster and produce higher-quality printouts, but they are also much more expensive (usually at least $1000). High-quality laser printers are also becoming available and may now be within the price range and capacity needs of the average person or small business.

The **magnetic media** (diskettes and tapes) used in input units are also useful for recording output from the computer. Cassette tape units cannot be recommended for any serious use; anything beyond the most rudimentary use of the computer will require at least two floppy-disk drives or perhaps a hard disk.

The remaining item of hardware that should be available in even a modest-sized operation is a **modem** (modulator-demodulator), which permits the computer to communicate over long-distance telephone lines with other computers. A modem converts electronic signals from the computer into signals that can be sent over a telephone line, and vice versa. The most common type is now the "direct-connect" modem, which can be plugged directly into the modular jack of the telephone.

Attached to a modem, the microcomputer can function as a terminal in accessing the programs and on-line data banks of large computer systems in distant locations and can **download** and retain software made available by other microcomputer users.

Putting It All Together

The final step in purchasing a microcomputer is selecting your software and hardware. The checklists in this chapter have been developed as guidelines for making your selection. The software evaluation is especially important for smooth operation of your microcomputer. The checklist on hardware will also help you ask the right questions.

CHECKLISTS FOR SOFTWARE EVALUATION

Three areas of software evaluation are important to consider before you make any purchases: (1) documentation or user materials need to be studied and evaluated; (2) vendor and dealer support should be investigated; and (3) the actual software applications programs or packages should be studied, used, and evaluated.

You can determine a composite score for each software program or package by adding the documentation, vendor support, and applications program subtotals on the following checklists and dividing that total by 3 to obtain a ranking on the basis of a total of 100 points.

Even if you don't use the point rankings, the checklists should be used to make sure you ask the appropriate questions. Software evaluation is an important task. Make sure that the software you buy does what you want it to do. The best way to assure this is to have hands-on experience with any software program you are considering.

Software Documentation Checklist

	No	Poor	Good	Excellent

1. An overview that gives an idea of the purpose and usefulness

   Score: (0) (1–3) (4–7) (8–10)

2. A tutorial section that will assist you in learning how to use the program

   Score: (0) (1–10) (11–20) (21–30)

3. A summary of commands

   Score: (0) (1–5) (6–10) (11–15)

4. A troubleshooting guide

   Score: (0) (1–10) (11–18) (19–25)

5. A technical section that describes how the software works

   Score: (0) (1–3) (4–7) (8–10)

6. A table of contents and an index

   Score: (0) (1–3) (4–7) (8–10)

Total (1 through 6) ___ ___ ___ ___

Vendor Support Checklist

	No	Poor	Good	Excellent
1. A telephone hotline to answer questions in the microcomputer start-up phase				
Score	(0)	(1–6)	(7–14)	(15–20)
2. A written warranty				
Score	(0)	(1–6)	(7–14)	(15–20)
3. Low-cost backup copies				
Score	(0)	(1–5)	(6–10)	(11–15)
4. Low-cost update as changes are made in software				
Score	(0)	(1–5)	(6–10)	(11–15)
5. On-farm training for program purchasers				
Score	(0)	(1–3)	(4–7)	(8–10)
6. Service for modifying programs				
Score	(0)	(1–3)	(4–7)	(8–10)
7. A newsletter and/or seminars available for the dealer				
Score	(0)	(1–3)	(4–7)	(8–10)
Total (1 through 7)				

Applications Program or Package Checklist

	No	Poor	Good	Excellent
1. Menus and prompts that are clear and concise				
Score	(0)	(1–3)	(4–7)	(8–10)
2. Prompts available on the screen to take care of errors				
Score	(0)	(1–5)	(6–10)	(11–15)
3. Accurate results				
Score	(0)	(1–5)	(6–10)	(11–15)
4. Operates at a desired speed				
Score	(0)	(1–3)	(4–7)	(8–10)
5. Adequate number of printout options				
Score	(0)	(1–3)	(4–7)	(8–10)
6. Easy to use				
Score	(0)	(1–5)	(6–10)	(11–15)
7. Can be modified for user by the vendor				
Score	(0)	(1–3)	(4–7)	(8–10)
8. User can modify the program				
Score	(0)	(1)	(2–3)	(4–5)
9. Useful on your farm or ranch				
Score	(0)	(1–3)	(4–7)	(8–10)
Total (1 through 9)				

CHECKLIST FOR HARDWARE EVALUATION

Following is a checklist that will help you compare types of microcomputer hardware. An estimation of system cost is found at the end of the checklist. This is an important part of the evaluation of a complete microcomputer system.

Hardware Comparison Checklist

General Characteristics

Brand _____ _____ _____

Amount of internal memory (K) _____ _____ _____

Clock speed _____ _____ _____

Type of processor chip:
    8-bit chips
      (MCS 6502, Zilog Z-80) _____ _____ _____

    8/16-bit chips (Intel 8088) _____ _____ _____

    16-bit chips (Intel 8086) _____ _____ _____

    16/32-bit chips (Intel 80286) _____ _____ _____

    32-bit chips (Motorola MC 68000) _____ _____ _____

Number of expansion slots _____ _____ _____

Maximum memory _____ _____ _____

Type of operating system:

    native DOS _____ _____ _____

    CP/M _____ _____ _____

    UNIX _____ _____ _____

    MS-DOS (PC-DOS) _____ _____ _____

UCSD p-System   _____ _____ _____

PICK DOS   _____ _____ _____

Type of printer interface connection:

    parallel   _____ _____ _____

    serial   _____ _____ _____

    both   _____ _____ _____

Screen Characteristics

Brand   _____ _____ _____

Features of monitor or screen:

    number of columns   _____ _____ _____

    number of lines   _____ _____ _____

    monochrome resolution   _____ _____ _____

    standard in pixels   _____ _____ _____

    monochrome with high-resolution board _____ _____ _____

    color with high-resolution board   _____ _____ _____

    number of colors   _____ _____ _____

    height, width, and depth   _____ _____ _____

    weight   _____ _____ _____

    screen size (diagonal)   _____ _____ _____

    type of screen (regular, liquid crystal,
       LED)   _____ _____ _____

    brightness controls available   _____ _____ _____

    contrast controls available   _____ _____ _____

    glare control available   _____ _____ _____

Any barrel distortion (yes or no)   _____ _____ _____

Your reaction to reading from screen for 10-
    minute period (poor, fair, good,   _____ _____ _____
    excellent)

Printer Characteristics

Brand   _____ _____ _____

Price   _____ _____ _____

Compatible with system unit
    (yes or no)   _____ _____ _____

Type of printer:

    dot-matrix   _____ _____ _____

    daisy wheel   _____ _____ _____

    thermal   _____ _____ _____

    other   _____ _____ _____

Kind of paper used:

    regular bond   _____ _____ _____

    special   _____ _____ _____

Width of carriage:

    9 1/2 inches   _____ _____ _____

    14 7/8 inches   _____ _____ _____

Type of paper feed:

    friction   _____ _____ _____

    tractor   _____ _____ _____

    sprocket or pin feed   _____ _____ _____

Paper format:

    cut or single sheet   _____ _____ _____

    roll or teletype   _____ _____ _____ fanfold   _____ _____ _____

Printing size variations available
    (number of fonts)   _____ _____ _____

Speed:

    maximum characters per second   _____ _____ _____

    average characters per second   _____ _____ _____

Unidirectional or bidirectional printing
    capability   _____ _____ _____

Noise (quiet, loud, very loud)   _____ _____ _____

Characters available:

    uppercase   _____ _____ _____

    lowercase   _____ _____ _____

    English alphabet   _____ _____ _____

    _____ alphabet   _____ _____ _____

    _____ alphabet   _____ _____ _____

    other characters   _____ _____ _____

Controls available:

    print on line and off line   _____ _____ _____

    form feed   _____ _____ _____

    line feed   _____ _____ _____

    form length   _____ _____ _____

    top of form   _____ _____ _____

    paper feed   _____ _____ _____

    line spacing   _____ _____ _____

Type of ribbon (cartridge or individual)   _____ _____ _____

Graphics capacity:

    dot addressable (pictures drawn with dots)   _____ _____ _____

    block addressable (crude graphics display formed with blocks or print)   _____ _____ _____

Costs of Component Parts

Brand   _____ _____ _____

Base price of hardware:

    central system unit   _____ _____ _____

    monitor or screen   _____ _____ _____

    keyboard   _____ _____ _____

    diskette drives   _____ _____ _____

    printer   _____ _____ _____

    modem   _____ _____ _____

    operating system   _____ _____ _____

    cable 1   _____ _____ _____

    cable 2   _____ _____ _____

    cable 3   _____ _____ _____

    display adapter   _____ _____ _____

    RS-232 post   _____ _____ _____

    expansion board or boards   _____ _____ _____

    other accessories   _____ _____ _____

    power surge protector   _____ _____ _____

    chips (memory)   _____ _____ _____

TOTAL HARDWARE COST $_____ $_____ $_____

Base price of software:

    Operating system type   _____ _____ _____

    Applications software

      farm accounting   _____ _____ _____

      word processing   _____ _____ _____

      spreadsheet   _____ _____ _____

      database management   _____ _____ _____

      communications (modem)   _____ _____ _____

      farm decision aids   _____ _____ _____

      project management   _____ _____ _____

      educational   _____ _____ _____

      games   _____ _____ _____

      _____ _____ _____

      _____ _____ _____

      other software   _____ _____ _____

TOTAL SOFTWARE COST $_____ $_____ $_____

TOTAL SOFTWARE AND HARDWARE
    COST $_____ $_____ $_____

SUMMARY

What will be the future use of microcomputers by farmers and ranchers? The future is always an extension of the present. Agricultural producers are categorized by size in terms of sales. In Chapter 1, emphasis was placed on the fact that about 10 to 15 percent of farmers with gross sales of more than $40,000 are using microcomputers in conducting their business operations. Most of the farmers or ranchers owning and operating microcomputers are using bookkeeping and accounting packages. By the year 2000, approximately 30 percent of the farm operators with sales over $40,000 are expected to use microcomputer hardware and software in their bookkeeping, inventory control, and planning or decision-making functions. Farm operators will be trained to use specific software applications packages and programs to carry out the planning, organizing, directing, coordinating, and controlling functions in their farm business. Farm operating decisions are becoming more and more complex with increased capital requirements, higher and changing interest rates, changing commodity prices, and more crucial market management choices.

A key factor in business survival is the ability to assemble, interpret, use, evaluate, and store the information that supports good decision making. Microcomputer hardware and software can effectively improve the quality and quantity of any information that you need to gather, interpret, and store for future use. Some of this information will continue to come from the various parts of the farm or ranch operation's enterprises, but outside information obtainable via modem from public and private sources will become increasingly important.

Market management decisions involve rather complete knowledge of weather supply conditions (affecting responses to government programs), demand conditions, and other geopolitical factors in the world. Each farm or ranch operator will need to determine the particular packages of written reports and publications, public source information, and/or commercial information needed. The microcomputer may be one way to obtain this information via telephone, radio, or satellite dish. Routine storage of such information for periodic review and use is an essential ingredient of a management information system. Specific decision-making aids developed on the microcomputer may also provide a basis for more effective pricing and selling of various agricultural commodities.

Information systems can supply sound information for tax management, business choices for profitable enterprises, family estate planning, market management, and personal decision making. Farm operators of the future will continue to have demands placed on them by the economic system that will require better management informational systems. Farm or ranch operators who take advantage of the opportunity to use microcomputers in developing sound information for making profitable decisions will continue to be competitive in American agriculture.

The microcomputer hardware and software evaluation material you have just read should provide a basis for helping you decide whether a microcomputer should be a part of your management information system. Even though you seek information from other sources before reaching a final conclusion, these materials should contribute to a better informed decision. If you decide that a microcomputer is for you, best wishes for its successful use in your farm or ranch operation.

VIEWS OF AGRICULTURAL SOFTWARE VENDORS

The Role of Computer Technology in Future Agriculture
by Norm A. Brown, President, Farm Business Software, Aledo, IL, June, 1989

In the 1980s, American agriculture focused its attention on getting its financial house in order. Computers played a key role in this process by automating and enhancing such traditional (and often mundane) tasks as accounting and budgeting.

If you are already organized and comfortable with financial records and computers, the 1990s will be an opportunity to dramatically increase efficiency and profitability. Just imagine an integrated, management information system that links together *all* facets of farm management including:

- Electronic identification and monitoring of each animal's health, production, and genetic capacity.
- On-the-go crop data recording, pest monitoring, and treatments tied to high-resolution field maps.
- Central monitoring and control of grain drying and storage and feed processing.
- To-the-second update of market, weather, and technical information.

All this information will be combined in an on-farm database that uses artificial intelligence to continuously optimize production practices and marketing strategies with the current conditions in the world and *your* operation.

Tomorrow you could be using computer technology to:

- Reduce costs.
- Increase production.
- Control more acres.
- Manage more employees.
- Produce better-quality crops and livestock.
- Generate more profit per enterprise.
- Receive top market prices.
- Negotiate the best financing terms.
- Free up more time for you and your family.

But only *if* you get off to the right start today.

What Role Will Computer Technology Play in the Future of Agriculture?
by Richard Moore, Farm Management Systems, R.R. 1, Box 42A, Manhattan, IL

What role will computer technology play in the future of agriculture? It's an interesting question. To begin to answer it, we might start by turning the clock back a few years and looking at how the computer has been involved up until now.

We really don't have to look back very far because microcomputers have been available for scarcely more than 10 years. Ten years ago, the average computer probably had 48K of memory, one or two floppy disk drives, and a text-only monochrome monitor. An average computer purchased today has at least 640K of memory, a 20- to 80-megabyte hard disk, and an enhanced color graphics monitor. And what about software? Ten years ago, there was very little software available, let alone software specific to agriculture. Today, there is a wide variety of very high-quality agricultural software as well as software for just about anything else. Computers are definitely a part of agriculture today and their adoption has taken place in a relatively short time frame.

So, where will we be 10 years from now? Computers are going to continue to do more in less time and take up less room. We are likely to see traditional electronic equipment like a planter monitor look more and more like a full-feature computer. As optical disk technology continues to advance, your office computer will virtually have libraries of information at its disposal. Instead of doing only one task at a time, expect your computer in the future to perform multiple tasks simultaneously. As the computer technology continues to advance, so will the software. Tomorrow's software will be easier to use, faster, and more powerful than the software we have today.

But, on the practical level, what does all of this mean for use today? Running a farm business is going to continue to get more and more complicated with less tolerance for error. Managing information will therefore become more critical and the computer will play an increasingly important role. If your farm uses a computer today, do your best to stay current with the technology. If a computer is not one of your management tools today, make it one and give yourself the competitive edge.

Glossary

(Boldface terms in the definitions are also defined in this glossary; → = see also.)

access: 1. Process of locating a certain item of information in the computer's **memory** and fetching it to the **CPU** for processing; 2. More broadly, the ability to locate and retrieve any type of information, such as a **file**, computerized data bank, etc.

acoustic coupler: A type of **modem** that converts electrical signals to and from the computer to sounds that can be transmitted through an ordinary telephone handset from and to another computer.

A/D (analog to digital conversion): Converting an **analog** signal (such as **output** from a measuring device) to **digital** representation. Examples: joystick, **mouse**, digitizer. → D/A

address space: The entire range of theoretically possible **memory** addresses in **RAM** that a computer can use to write information to and read information from.

algorithm: A sequence of steps such that, when correctly completed, will yield the solution to a problem.

ALU (arithmetic-logical unit): Component of the **CPU** where information is represented electronically in the **registers** while arithmetic and logical operations are carried out on it.

analog: Any device that works on the principle of measuring some continuously variable physical quantity, such as how far a **mouse** has been moved on a desktop. → A/D, digital

ANSI (American National Standards Institute): A private organization in New York City that publishes proposed standards for many products, including computer **languages** such as **BASIC** and **Fortran**. Adherence to these standards is entirely voluntary.

applications program or package: A specialized **program**, or set of programs, designed for a particular type of task, such as **database, graphics,** or **desktop publishing** uses.

architecture: The way in which the components of a computer **system** are organized and interconnected.

artificial intelligence (AI): The study of using computers to simulate human thinking processes such as learning, reasoning, and problem-solving.

ASCII (American Standard Code for Information Interchange): A standard 7-**bit** code for data transmission in which every **character** has a unique **binary** representation.

assembler: A programming **language translator** that converts code with a close relationship to the **hardware** organization of the computer into executable **binary**; assembly language is thus very flexible and efficient, but correspondingly complex and "technical" (esoteric).

backup: An extra copy of a data set (**file, disk**), made as a precaution against accidental loss, damage, or erasure.

bar code: A form of optically readable information encoding now widely used on groceries and other small retail items. → OCR

barrel distortion: A type of distortion on the **screen** of a **monitor** in which a rectangle will appear to have bowed sides (like a barrel).

BASIC (Beginner's All-purpose Symbolic Instruction Code): A relatively simple and easily learned computer **language** developed at Dartmouth College in the early 1960s. Though scorned by computer professionals, it is still perhaps the most widely used language for programming personal or home computers. → Pascal, C

baud: A unit of the rate of information transmission; a 1200 baud connection transmits about 120 **characters** per second from one device to another. (Named after a Frenchman, Baudot, inventor of an early signalling code.)

binary: Refers to the base 2 system of representing information, in which every item appears as a sequence of 1's and 0's.

bit: **Binary** digit; the 1 or 0 used to represent the smallest possible unit of information.

bootstrap: The initial information **loading** that takes place when a computer is turned on and prepares itself for work; this enables it to read in and respond to further instructions. Also called "booting up."

bpi or BPI (**bits** or **bytes** per inch): The **density** with which information is encoded on a recording medium such as magnetic **disk** or tape. → magnetic medium

buffer: A special area of **memory**, often used to compensate for the differing speeds of operation of **hardware** components of a computing **system**.

bug: A **software** error or, sometimes, **hardware** malfunction. → debugging

bus: Circuit or pathway along which electronically coded information of a certain type travels inside the computer; there is usually an instruction bus, an address bus, and a data bus.

byte: A standard-length sequence of **bits** (usually 8) used to represent a single **character** (letter or numeral); thus also a unit of **memory** capacity. → kilobyte, megabyte

C: A very concise **programming language** now widely used for writing **operating systems** and **applications packages** because of its compactness and efficiency.

CAD (Computer Assisted Design): A type of advanced **graphics** program that assists the design of industrial products.

character: Symbol that can be processed by the computer; may be alphabetic (upper- and lowercase letters A–Z and a–z), numeric (digits 0–9), or special (blank, period/decimal point, punctuation marks, dollar sign, percent sign, etc.).

chip: 1. **Microprocessor** in which the computer's transformations of information are done, containing the

CPU and **control unit.** 2. More generally, any **microprocessor, memory** module, or other integrated circuit element.

clock: Component of the **control unit** that emits the high-frequency pulses used to synchronize all other operations of the computer.

COBOL (COmmon Business Oriented Language): The most widely used **programming language** in business data processing. → **Fortran.**

compatible: 1. Refers to components that work together within a single computing **system** (for example, a computer and its printer must be compatible). 2. Refers to a computer designed to imitate closely the operating characteristics of another brand of computer, primarily so that it can use **software** developed for the other brand. → workalike

compiler: A programming **language translator** that converts a "higher-level" computer language (such as **Fortran** or **Pascal**) into **binary** form, such that it can be **executed** ("run") by the computer.

computer: An information-processing device or machine that may be of **analog** or **digital** type and may be mechanical, electromechanical, or electronic in mode of operation.

control unit: The component of a computer's **CPU** that coordinates all of its functions and processes. → ALU, clock

CPM (Critical Path Method): A quantitative method for the analysis of complex projects, indicating which tasks and subtasks are crucial to their quickest completion. → PERT

CP/M (Control Program/Microcomputer): A once widely used **operating system** for **microcomputers.** A proprietary product of Digital Research, Inc., of Pacific Grove, California. → DOS

CPS (characters per second): A unit of the speed of operation of a printer or data transmission device.

CPU (central processing unit): The "brain" of the computer, which includes the **control unit** and **arithmetic-logical unit.**

CRT (cathode ray tube): The television-like display device used as a **monitor** on most computers. → VDT.

cursor: The indicator light on the **monitor screen** that shows "where you are," or where the next thing typed will appear.

D/A: Digital to analog conversion, or converting **digital output** from the computer to movements of some physical device such as the pen of a **plotter** or a **robot** arm. → A/D.

database: An organized collection of interrelated information about a particular field, topic, or set of facts.

database management system (DBMS): An **applications package** designed to keep records in a database, to make it easy to create, modify, and delete records and locate and view selected records.

debugging: The process of locating, identifying, and correcting errors, as in a **program**, or correcting an equipment malfunction. → bug

dedicated computer: A computer designed or reserved for one specific purpose, such as **word processing** or statistical analysis.

default: The "plain vanilla" option, which is assumed if you fail to specify otherwise. It usually represents the **system** designer's guess at what most users would want most of the time.

density: The compactness with which information is recorded on the surface of a **magnetic medium.** Floppy diskettes may be double density or high density.

desktop publishing: Use of a **microcomputer** to lay out and produce high-quality printed materials with a variety of headings, multiple columns, diagrams, and photographs, etc. → word processing

digital: Any device that works on the fundamental principle of counting discrete objects or events, then recombining these counts in various ways. An electronic digital computer basically counts electronic pulses emitted by the **clock** in its **CPU.** → analog, A/D

direct-connect modem: A **modem** that plugs directly into the modular jack or the telephone outlet, making an **acoustic coupler** unnecessary.

disk or disc: A flat, circular, platelike device with a magnetic coating on which information can be encoded and stored as patterns of tiny magnetized spots. May be "floppy" (flexible) or "hard" (rigid), removable or "fixed" in place. → magnetic medium

documentation: All the written (usually published) material that describes an item of **software** or **hardware** and tells how to use it. The documentation may also be internal to a **program** if placed on REMARK or COMMENT lines within the **source code** listing.

DOS (Disk Operating System): 1. Any **operating system** that emphasizes the use of **disks** to store **files** of information. 2. In particular, the very widely used MS-DOS or PC-DOS operating system developed for **PC compatibles** by Microsoft, Inc.

dot-matrix printer: **Printer** that forms characters on the paper by activating various patterns of "needles" (tiny rods) within a rectangular grid arrangement.

download: The process of accessing **files** on a remote computer and copying them into the file system of one's own computer. → upload, modem

drive: An **input** or **output** device that reads from or writes on **magnetic media**. → disk

dumb terminal: A **terminal** that can send and receive only single lines of information from a remote computer, without the capacity to **upload** or **download files**. → modem, network

EBCDIC (Extended Binary Coded Decimal Interchange Code): An 8-**bit** code for information representation widely used in IBM computers. → ASCII

echo: The appearance of each **character** typed at the **keyboard** on the **monitor screen**, enabling the user to see what is being entered.

editor: A **software package** used to examine and modify the contents of **text files**, but usually lacking the reformatting and **word wrap** features of a **word processor**.

execute: To "run" a **program** with the computer responding to

the **binary** version of the information processing steps
specified by the instructions in the program.

file: A logically related collection of information; may contain
text (a **program** or natural-language discourse), a numeric
data set, or other kinds of information.

floating point operation (FLOP): A mathematical operation
involving decimal numbers; i.e., 23.4 × 0.00731. →
megaflop

floppy disk or diskette: A flexible (usually plastic) **disk** used
to record information in most small computer systems. →
magnetic medium, disk, density, DOS, byte

font: A particular design or shape and size of lettering or of
a typeface for printing.

format: The exact, detailed arrangement into which information
is organized. → sector, track

formed-character printer: **Printer** that uses character shapes
already sculpted into some base in its printing element, as
in a typewriter. → impact printer, dot-matrix printer

FORTH: Programming **language** for **microcomputers**, not as
widely used as **BASIC** or **Pascal**, but considered preferable
by its enthusiasts because of its conciseness and elegance.

Fortran or FORTRAN (FORmula TRANslator): One of the
oldest and most widely used programming languages,
especially in mathematical, scientific, and engineering
applications. → COBOL

graphics: Use of the computer to produce pictures or images
such as diagrams, maps, charts, graphs, patterns, slides,
animation, or the like. → high-resolution graphics, pixel,
jaggies

hard copy: **Printout** that is displayed on paper or some other
permanent medium.

hardware: All the physical, tangible components of a computer
system.

high-resolution graphics: **Graphics** with a high degree of detail
in the representation, perhaps by use of a large number of

pixels per inch.

impact printer: A printer that works on the principle of striking at the paper through an inked ribbon, like a typewriter. → nonimpact printer

input: Information that is put into the computer as its "raw material" for processing; may be numbers, words (text), images, tables of data, etc. Also the process of entering this information.

input form: A paper form, usually containing blanks to be filled in, used to organize information that will be called for during the running of an **applications package**.

integrated package: An **applications package** that combines several widely used types of applications, usually including at least **word processing**, **spreadsheet**, **database**, and perhaps graphics and communications features; the compatibility of **files** among these various types of functions is the major advantage.

interface: The point of contact between two parts of a computer system; for example, between a person and the computer, or between two **hardware** devices, as in data transfer through a **parallel** or **serial port**.

interpreter: A computer **language translator** that decodes and **executes** a **program** essentially line by line; it is thus relatively slow and inefficient compared with a **compiler**.

jaggies: The small breaks in **graphic output** caused by rows of **pixels** that do not lie along a perfectly smooth curve.

joystick: An **A/D** device that allows the control of the **cursor** or some part of the image on the computer **screen** by pressing a small lever. → mouse

justify: To align the margins of columns of **printouts** so that they appear in a straight vertical line. Typewritten text is usually left-justified with a "ragged right" margin; typesetters and **word processors** are able to right-justify as well.

K: See **kilobyte**.

keyboard: The typewriter-like **input** device used to enter **character** information into the computer.

kilobyte (K or KB): One thousand (actually 2**10 = 1,024) **bytes**. A frequently used unit of information storage or **memory** capacity. → megabyte

language: A system of encoding instructions to the computer. Each language has its own rules of **syntax** and characteristic strengths and weaknesses for particular types of jobs. Must be decoded and translated into **executable binary** by a **translator** (**assembler**, **interpreter**, or **compiler**). Examples: **BASIC**, **Pascal**, **C**, Lotus 1-2-3 macros, dBASE command files. → source code

LCD (liquid crystal display): A type of **output** display device in which **characters** are typically formed by black lines on a gray background.

LED (light emitting diode): A type of **output** display device in which **characters** are formed by glowing line segments.

load: To copy information from a storage device or unit into the computer's **memory**, where it is available for processing.

LOGO: A programming **language** especially suited to the production of simple **graphics** by children. → BASIC

M: See **megabyte**.

magnetic medium: A form of information storage and transportation in which the information is encoded and recorded as patterns of magnetized dots on a **disk** or tape coated with iron oxide.

mainframe: A large computer capable of being accessed simultaneously by many **terminals** or **workstations** in **timesharing** mode.

megabyte (M or MB): One million (actually 2**20) **bytes** of information storage capacity. Frequently called a "meg." → kilobyte

megaflop: One million **floating point operations** per second; a unit of computational speed used in comparing **mainframes** or **supercomputers**. A CDC Cyber 205 has been operated (briefly) at an execution speed of almost 800 megaflops. The Japanese are said to be building computers that operate "up to" 1,300 megaflops (i.e., 1.3 gigaflops).

memory (or storage): The subunit of the computer where information is retained to make it available for processing by the **CPU**. The size (information storage volume) of memory is measured in **kilobytes** (K) and **megabytes** (M).

menu: Display of a list of options available at a given point in **program** execution. The user is then expected to choose one of the displayed options as the next operation of the program or package. **Software** controlled in this way is said to be "menu-driven."

MICR (magnetic ink character recognition): A machine-readable type of information representation using ink containing magnetic particles that allow recognition of **character** shapes by a special **input** device (as along the bottom of most preprinted bank checks).

microcomputer: A general-purpose computer small and inexpensive enough to be owned by an average individual; this is sometimes taken to mean that the total hardware cost is under $10,000.

microprocessor: The "brain" or **CPU** of a small computing system, made by fabricating thousands of circuit elements (transistors) on the surface of a silicon wafer in what is called a "large-scale integrated (LSI) circuit."

minicomputer: A computer somewhat larger, more powerful, and more expensive than a **microcomputer**, such as would be used by a larger business.

MIPS (millions of instructions per second): A unit of computational speed. Note that these are usually assembly **language** instructions, many of which may be required to carry out a single arithmetic or **floating point operation**. → megaflops

modem: MOdulator-DEModulator, the device that converts electrical signals to and from the computer into a form in which they can be transmitted across telephone lines. One type uses the telephone handset with an **acoustic coupler**.

monitor: The **screen** that displays **input/output** information and messages from the computer to the user. → CRT, VDT, scrolling

mouse: Hand-held **A/D** device that controls the movement of a pointer on the **monitor** screen and having one to three buttons on top enabling the user to **access** various features. Similar in function to a joystick, trackball, or thumbwheels.

network: The process or product of interconnecting computers so that they can communicate, share **files** and **peripherals**, etc.

nonimpact printer: A printer that does not use the impact principle; for example, a thermal, ink-jet, and laser printer. → impact printer

OCR (optical character recognition): **Input** of information to the computer by use of a device that allows it to recognize the shapes of **characters**. → bar code

operating system (OS): A **software** package that manipulates **files** and allocates **system** resources to various computing functions such as copying, printing, editing, and executing files. Sometimes called the monitor, supervisor, or master control **program**. Examples: **DOS, CP/M, UNIX.**

output: Processed information coming out of the computer; may be in the form of printed numbers, **formatted** text, **graphics** images, or even the movement of some mechanical device such as a **robot** arm. → D/A, hard copy, plotter

package: A specialized **program**, or interrelated set of programs, used for solving problems of a certain type: **spreadsheet**, statistical, accounting, etc.

parallel: Suggests two or more processes going on side by side simultaneously, such as computation or data transfer. A parallel **port** sends **bytes** out "sideways" along eight separate paths. → serial

Pascal: A structured programming **language** developed for the purpose of teaching good programming practice, now

widely used in instruction and systems programming. →
BASIC, C

PC (Personal Computer): 1. Any **microcomputer**. 2. More
narrowly, the series of personal computers sold by IBM in
1981–1988 and the many **compatibles** made by others.

peripheral: Any **hardware** component of a computer **system**
other than the computer itself, especially **input** and **output**
devices. These can be attached to a **CPU** to increase its
usefulness. → monitor, printer, plotter

PERT (Program Evaluation and Review Technique): A method
for the planning and control of complex projects, primarily
by use of network diagrams to represent interdependence
of subprojects. → CPM

pixel: A single dot or point in a digitized **graphics**
representation, such as a **monitor screen**.

plotter: Device for producing **hard copy graphics output**; one
or more pens are moved under control of the computer
over a large sheet of paper mounted on a drum or flat
bed.

port: A point of access to the circuits and information flow
inside a computer.

prettyprinting: The process or product of attempting to
produce attractive, readable, easily understandable
printouts.

printer: Device for producing a **hard copy** record of
human-readable **output**, as of computational results or
formatted text. → impact printer, nonimpact printer

printout: The human-readable result of computation or
processing, as displayed on a **monitor** or on **hard copy**.

program: A detailed list of instructions to the computer to
carry out the steps of some information processing
procedure, normally written in a standard programming
language such as **BASIC**, **Pascal**, or **C**.

prompt: A signal from the **system** or a **program** that it is ready
to receive and process the next instruction or item of
input.

protocol: The common organization of transmitted information
that allows **networked** computers to exchange messages and

files.

public domain: Said of **software** that is freely copyable and usable because nobody claims ownership or property rights in it. → shareware

RAM (Random Access Memory): **Memory** that can both be read from and written to while the computer is running. Information stored here is volatile, meaning that it is lost when the computer is turned off.

registers: The portion of an **ALU** where **binary** information is represented electronically while logical and mathematical transformations are carried out on it.

robot: A computer-controlled machine for manipulating objects. → D/A, output

ROM (Read-Only Memory): **Memory** that can be read from at any time but not written to by action of the computer. Often used for permanent storage of system **utility programs** such as the **operating system** or BASIC **interpreter**.

save: To transfer (copy) information from the computer's **memory** onto a storage medium (such as magnetic **disk** or tape) for retention until a later date. → magnetic medium

scanner: An **input** device that converts a **graphic** image or a typed or printed page into **digital** form so that it may be stored and manipulated within the computer.

screen: The television-like area used to display **input** and **output** information on a computer **monitor**.

scrolling: The movement of lines of text up the **monitor screen** (and off the top) as additional information is placed on the bottom line by the user (**input**) or by the computer (**output**).

sector: Wedge-shaped portion of a **disk**, used by the **operating system** to organize blocks of information for storage on the disk. → magnetic medium, track

serial: Suggests a sequential process, or one event taking place after another. A serial **port** sends **bytes** end-to-end to another device, as if along a single wire.

shareware: A type of **software** distribution in which the developer encourages the widespread copying of his or her **program** and the voluntary payment of a fee (usually $5 to $100) by those who make use of it. → public domain

simulation: Use of the computer to artificially depict hypothetical changing situations (such as market movements or weather fluctuations) that are difficult, expensive, or impossible to actually observe or experiment on.

software: The information that directs a computer system to perform some function, such as a **program** or **applications package**. Usually encoded on a recording medium such as magnetic **disk** or tape.

source code: A complete and corrected written listing of the instructions that make up a **program**.

spreadsheet: An **applications package** that allows the entry and manipulation of information in a row and column **format** similar to an accountant's large ledger sheet.

supercomputer: One of the fastest and most powerful computers available at any given point in time. Examples: Cyber 205 (Control Data Corp.) and the Cray X-MP (Cray Research, Inc.). A Cyber 205 has been operated briefly at almost 800 **megaflops**.

support: To assume responsibility for making something in a computing **system** operate or work properly.

syntax: The rules or "laws" according to which expressions (such as instructions to the computer) must be formed, similar to the rules of grammar or spelling in English.

system: The totality of all the components that make a computer operate, considered as a unit or interrelated whole: **hardware**, **software**, and sometimes the people (programmers, engineers, users, etc.) who interact with them.

template: A "fill in the blanks" form that can be filled in with information as needed, such as a generic **spreadsheet** or **word processing format**.

terminal: A **peripheral** device that accepts **input** from a

keyboard and displays **output** on a **monitor** or printing device.

text: A data set encoded in a way that can be read directly by users. A **text file** can be alphabetic, numeric, or a combination of the two (alphanumeric or "alphameric"). Usually contrasted with **binary**.

thimble: A type of formed-character printing element used in some **impact printers**, so called because of its shape.

timesharing: A type of **operating system** in which a number of users working at **terminals** can **access** and work on a single large computer simultaneously. → mainframe

touch-switch keyboard: A once popular type of **keyboard** in which information is entered by pressing spots on a membrane instead of pushing down keys.

track: One of the concentric circles on a disk (or straight line paths along a tape) along which information is recorded. → magnetic medium, disk, sector, format

translator: A **software package** that converts **source code** in some standard programming **language** (such as **BASIC**, **Pascal**, **C**, etc.) into executable **binary** code. May be an **interpreter**, **compiler**, or **assembler**.

UNIX: A type of concise **operating system** used to operate **mainframe** computers in **timesharing** mode and on some of the more powerful **minicomputers** and **workstations**.

upload: The process of copying **files** from the file system of one's own computer into the files of a remote computer with which one can communicate. → download, modem

user-friendly: Term for **hardware** and **software** that most people, even those with no computer training, find relatively easy to learn and use. → menu, documentation

utility program: A general-purpose **program** intended to improve the operation of a computer **system**; contrasted with an **applications program**. Example: an **editor** or editing program for **text** or **ASCII files**.

vaccine: a **program** intended to counteract or eliminate a certain **virus**.

vendor support: The **support** provided for a computer **system**, or some component of it, by whoever sold it to the user.

VDT (video display terminal): A combination **keyboard** and **CRT monitor** used for **input** to and **output** from a computer **system**.

virus: a **program** that has the capacity to reproduce itself and insert itself into other programs. → vaccine

window: A rectangular portion of the computer **screen** in which information of a certain kind is presented; the user may see several windows at once or switch from one to another.

workalike: A computer designed to work as closely as is legally possible like another (usually very popular) brand, and especially to use the same **software**. An (almost) completely **compatible** workalike is sometimes called a clone.

workstation: A combination **microcomputer** and **terminal** that can operate independently in stand-alone mode and can also send **files** to a host **mainframe** elsewhere when greater computational power is required. Usually also has advanced **graphics** capability.

word processing: Use of a specialized **applications package** to create, interactively modify, and print out any sort of textual material or document. Replaces a typewriter but is not as elaborate as **desktop publishing**. → format, printer, font

word processor: **Applications package** that **formats** and prints out textual material in an attractive way; also certain specialized or **dedicated computers** used only for this type of **applications program.**

word wrap: A feature of **word processing** and other **packages** in which **text** lines too long to be displayed are automatically broken and continued on the next line.

Index